IT'S DEBATABLE!

USING SOCIOSCIENTIFIC ISSUES TO DEVELOP SCIENTIFIC LITERACY

K–12

USING SOCIOSCIENTIFIC ISSUES TO DEVELOP SCIENTIFIC LITERACY K–12

Dana L. Zeidler and Sami Kahn

Arlington, Virginia

Claire Reinburg, Director
Wendy Rubin, Managing Editor
Andrew Cooke, Senior Editor
Amanda O'Brien, Associate Editor
Amy America, Book Acquisitions Coordinator

ART AND DESIGN
Will Thomas Jr., Director
Joe Butera, Senior Graphic Designer, cover and interior design

PRINTING AND PRODUCTION
Catherine Lorrain, Director

NATIONAL SCIENCE TEACHERS ASSOCIATION
David L. Evans, Executive Director
David Beacom, Publisher

1840 Wilson Blvd., Arlington, VA 22201
www.nsta.org/store
For customer service inquiries, please call 800-277-5300.

17 16 15 14 5 4 3 2

NSTA is committed to publishing material that promotes the best in inquiry-based science education. However, conditions of actual use may vary, and the safety procedures and practices described in this book are intended to serve only as a guide. Additional precautionary measures may be required. NSTA and the authors do not warrant or represent that the procedures and practices in this book meet any safety code or standard of federal, state, or local regulations. NSTA and the authors disclaim any liability for personal injury or damage to property arising out of or relating to the use of this book, including any of the recommendations, instructions, or materials contained therein.

Library of Congress Cataloging-in-Publication Data
Zeidler, Dana L. (Lewis) author.
It's debatable! : using socioscientific issues to develop scientific literacy, K-12 / Dana L. Zeidler and Sami Kahn.
pages cm
Includes index.
ISBN 978-1-938946-00-4
1. Science—Social aspects—Study and teaching—United States. 2. Technology—Social aspects—Study and teaching—United States. 3. Curriculum planning—United States. I. Kahn, Sami, author. II. Title.
Q175.5.Z38 2014
507.1'173—dc23
2013042407

Cataloging-in-Publication data are also available from the Library of Congress for the e-book.
e-LCCN: 2013044019
e-ISBN: 978-1-938946-64-6

CONTENTS

Unit 1

Food Fight

Should schools charge a "fat tax" for unhealthy foods?

Unit 2

Animals at Work

SPEED
LIMIT
25
55

High School Level

Unit 7

"Pharma's" Market

Should prescription drugs be advertised directly to consumers?

Contributors

The authors are grateful for the expert contributions of the following outstanding educators:

"Voices From the Field": Teacher Perspectives

Scott Applebaum
Palm Harbor University High School
Palm Harbor, FL

Hyunsook Chang
Kuksabong Middle School
Seoul, South Korea

Thomas Dolan
Pride Elementary School
Tampa, FL

Kisoon Lee
Sinam Middle School
Seoul, South Korea

Unit Plan Contributors

Brian Brooks
University of South Florida
Tampa, FL

Michael Caponaro
University of South Florida
Tampa, FL

Kory Bennett
University of South Florida
Tampa, FL

Jessica Croghan-Ingraham
University of South Florida
Tampa, FL

Christina Cullen
University of South Florida
Tampa, FL

Thomas Dolan
Pride Elementary School
Tampa, FL

Katie Frost
University of South Florida
Tampa, FL

Bryan Kelly
University of South Florida
Tampa, FL

Daniel Majchrzak
University of South Florida
Tampa, FL

Andrea Churco Marshall
Hillsborough County School District
Florida

Lisa Clautti Mistovich
University of South Florida
Tampa, FL

Tammy Modica
University of South Florida
Tampa, FL

Crystal Nance
University of South Florida
Tampa, FL

Ashley Schumacher
University of South Florida
Tampa, FL

Hayley Sweet
Florida College Academy
Tampa, FL

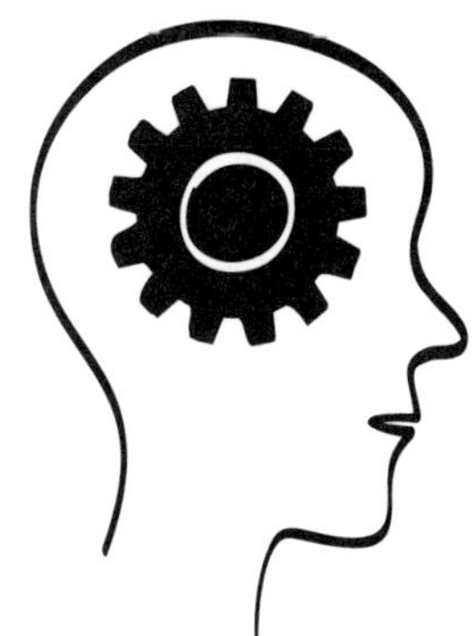

About the Authors

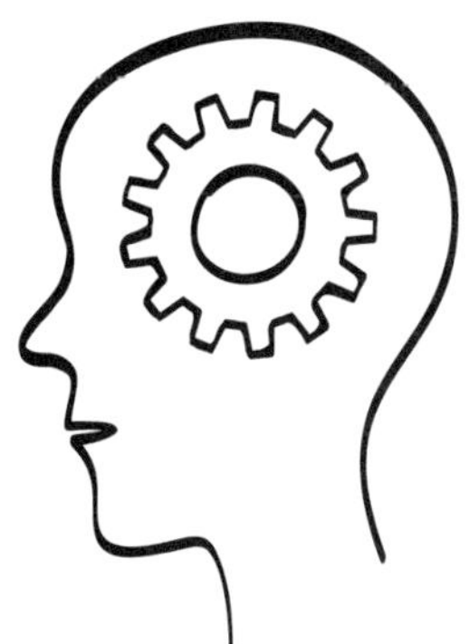

Dana L. Zeidler (first author) earned his PhD from Syracuse University. His research program incorporates aspects of socioscientific issues as a means to facilitate scientific literacy. His work has attracted international attention and is cited widely both within and external to the field of science education. His line of extensive research can be found in numerous journal articles, book chapters, keynote addresses and international conference proceedings. He works closely with doctoral students and other leaders in the science education community. He is a Professor and Program Coordinator of Science Education at the University of South Florida, Tampa Bay. Dana has long-standing ties to the science education community including working closely with other leaders, faculty, and graduate students. Some of his most recent honors include:

- President of National Association for Research in Science Teaching (NARST): A worldwide organization for improving science teaching and learning through research, 2010–2011
- Recipient for the Outstanding Mentor Award (2008), Association for Science Teacher Education (ASTE)
- Executive Board of Directors, NARST, 2006–2009
- At Large Board of Directors, ASTE (2008–2011)
- Series Editor for *Contemporary Trends and Issues in Science Education*, Springer: Dordrecht, The Netherlands (2008 to present)
- Distinguished Visiting Professor of Science Education, Ewha Womans University, Seoul, South Korea (2012–2013)
- Honorary Professor of Science and Environmental Studies, The Hong Kong Institute of Education, The University of Hong Kong, China (2013–2016)
- Attained level of Master (7th-degree Black Belt) from the Okinawa Isshinryu Karate Association, Okinawa, Japan (1982 to present)

Sami Kahn (second author) is a 26-year veteran science educator with extensive experience in classroom teaching, professional development, and curriculum development. Currently serving as a Presidential Doctoral Fellow in Science Education at the University of South Florida, she has authored numerous journal articles, including several in *Science and Children*, and has coauthored three books on enhancing scientific inquiry experiences for children and adults. She has served as an invited and keynote speaker at several state and national conferences, and most recently at an international STEM conference in Thailand. She is particularly known for her work in ensuring quality science opportunities for all children, including those with disabilities. In that capacity, she has served as president of Science Education for Students with Disabilities (SESD), an NSTA Associated Group dedicated to inclusive science practices, chair of the National Science Teachers Association's Special Needs Advisory Board, and chair of the national awards committee for the Scadden Science Teaching Award. She also had the honor of serving as a delegate to the National Congress on Science Education, from which she was elected to represent the Congress as an NSTA national convention planning committee member. Ms. Kahn has successfully taught grades Kindergarten through college, as well as inservice professionals, and has won numerous awards for outstanding science teaching. She holds an MS in ecology and evolutionary biology and a JD with an emphasis in environmental law from Rutgers University. Prior to coming to University of South Florida, she most recently served as lower school science coordinator/teacher and K–12 science department chair at Collegiate School in New York City.

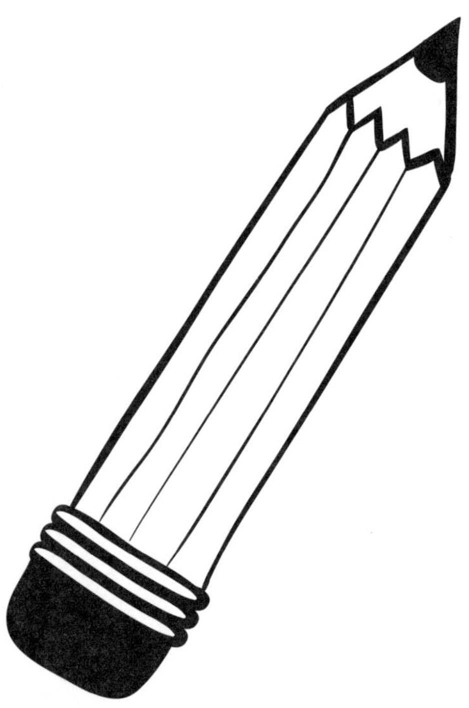

A Prelude

"Monday Morning"

Ms. C. left the school office in a hurry, hastily grabbing her mail before heading to her first period biology class. Mondays were always a challenge, but today seemed especially so. It wasn't even 8:00 and she had already been involved in a heated discussion of the new accountability measures being implemented in her school district; more tests, more requirements to meet Common Core, and more pressure to make sure her students made Adequate Yearly Progress or her performance evaluations would be on the line. "When did teaching become so stressful?" she thought. It was never an easy job, but the last few years felt especially weighty … less focused on the subject and students she loved, less creative, and definitely more stressful.

She took a deep breath as she entered her classroom. "Good morning!" she said with a somewhat forced smile. Almost immediately, she sensed something was different today… her ninth graders, usually sleepy on Monday mornings, were talkative and animated. The energy in the room was palpable.

"Hey, Ms. C, we were just arguing about the vaccine question from last week. I still think it's wrong to make teenagers take vaccines!" said Alex.

"But they protect us!" exclaimed Janelle.

"That's what the drug companies want you to think. What's your evidence?" asked Vincent.

Ms. C's ears pricked up as she was shocked to hear her students suddenly engaged in an impromptu discussion of … science!

"My group read an article about how the vaccine protects against cervical cancer, so it's a good thing!" Janelle retorted.

"Yeah, but what was the source? Did you evaluate the source?" pressed Vincent.

Ms. C. smiled in awe. She didn't know if the discussion on evaluating sources of evidence would sink in. It didn't sound like the most exciting topic when she read about it, and yet, the students really enjoyed evaluating different websites and articles using the rubric she had provided. Maybe this new curricular approach was working.

"But our group found an article that said that vaccines were bad for you. There are side effects … even death!" replied Alex.

"Yeah, that's the problem with science. There are always different reports. You never know what to trust!" added Crystal, in an exasperated tone.

"But that doesn't mean you can't trust it. Remember the whole nature of science thing? There's always new information," added Miguel.

Ms. C. felt a tingling feeling she hadn't felt in years: that combination of pride and excitement that comes with knowing you've impacted your students' lives.

"I don't think it matters whether the vaccine helps or not. No one has the right to make me take a vaccine if I don't want it. It's my body! And I talked about this with my parents this weekend and they agree!" exclaimed Karla.

Ms. C. couldn't believe her ears. She had never heard a peep from Karla, her quietest student. Yet today Karla was taking a stand, and she had clearly been talking (and thinking) about it over the weekend.

"Maybe this new SSI approach I'm trying is making a difference," Ms. C. thought. In fact, her friend Mr. Alvarez, a Language Arts teacher, did mention that he heard some of Ms. C's students discussing the vaccine argument in his class last Friday … science seemed to be spilling over into other parts of the day. And yet, Ms. C. really hadn't made a huge change in her teaching. She had just decided to tweak her already-existing curriculum to include a few extra lessons that put the science content into a personal context that really motivated and challenged her students … and her.

Ms. C hated to interrupt the students, who at this point were engaged in a full debate. "Let's take a look at what we've learned so far and see what we still need to discover!" she said.

It was going to be a good day … and a good year!

Part 1

Introduction and Background to Socioscientific Issues

Why Socioscientific Issues?

It is nearly impossible for a day to go by without hearing or reading some discussion of scientific issues. From debates over stem cell research to whether fluoride should be added to drinking water, from genetically modified crops to so-called "fat taxes" to discourage consumption of fatty foods and drinks, every day citizens are deluged with questions, controversies, and often conflicting "facts" about issues that require their deliberation. Television commercials for prescription drugs, and layperson's blogs discussing scientific research findings, for example, remind us that while earlier generations relied on "experts" including scientists and doctors to dictate policies and protocols, today's generation demands an active voice in decision making about their health, environment, technology, and their world. Unquestionably, their technologically and scientifically advanced world has become so connected that every decision has potential for global impact. From large-scale issues such as climate change, deforestation, and nuclear energy, to more local concerns such as bicycle helmet regulations and beach restoration, scientific issues increasingly permeate everyday life.

It is clear that today's students need to be "science literate" in order to contend with these issues, but what exactly does that mean? More than simply knowing the content included in our science standards, our students need to be able to synthesize, analyze, and deliberate over complex socioscientific issues (SSI), which will demand their attention, their input, and perhaps, their vote. Today's students—tomorrow's informed citizens—must have some modicum of *functional scientific literacy*; that is, the ability to apply scientific reasoning to real-world scenarios, a daunting task that requires recognition of the scientific issues, research into and analysis of information, thoughtful deliberation, and an understanding of the moral and ethical implications of their decisions. It also necessitates an understanding of the tentativeness of scientific findings, the realization that scientists discuss and disagree, and of course, the ability to become comfortable with the discomfort of uncertainty. In short, functional scientific literacy requires an understanding of the nature of science and the skills necessary to think both scientifically and ethically about everyday issues.

Emphasizing functional scientific literacy in our teaching reflects more than a desire to help students meet contemporary societal demands; it is an educational policy mandate. The *Next Generation Science Standards* (*NGSS*) require students to apply scientific understandings to everyday personal and societal contexts while the movement toward *Common Core State Standards* emphasizes interdisciplinary connections between science, language arts, and math. As if making science relevant, interconnected, and applicable to daily life weren't challenging enough, teachers are also obliged, under the No Child Left Behind Act, to ensure that all of their students are performing satisfactorily on standardized assessments. Clearly, teachers need an approach that garners the interests of an ever more diverse student population, emphasizes standards-based science learning, and fosters students' abilities to *use* science in addition to comprehending it. An approach that places science in a context that is meaningful to students' lives and requires students to consider the *moral and ethical implications* of their decisions not only prepares students for informed citizenship, but also helps teachers to meet the myriad demands being placed upon them in an age of accountability and assessment.

The idea of teaching "issues-based" curricula in science is not a new concept. Science, Technology, and Society (STS) has been a staple of many school programs and was an insightful effort at acknowledging the necessity of bringing science into everyday decisions. However, by simply presenting scientific concepts in the context of current issues, some important aspects of the nature of science, as well as the nature of people, can get lost. Take for example, the issue of childhood vaccinations. It is possible to introduce the study of the immune system by sharing with students the huge successes of vaccinations in terms of reduced mortality and disability from debilitating diseases. But what about the controversies? Do we steer clear of the questions about safety, efficacy, and parental choice because they are too "messy" and seem at odds with a traditional curriculum? Unfortunately, many well-meaning educators do steer clear of the controversy for fear that it "muddies the waters" for students and might also get the class off track from the "prescribed" curriculum. However, there is increasing evidence that the controversy *is* the central component of the curriculum, for it is necessary so our students will become scientifically literate members of society through practice in research, analysis, and argumentation of these controversies. Moreover, contrary to the manner in which the history of science has been traditionally taught, science doesn't progress neatly and linearly through "sterile" topics, but rather, through a series of "messy" questions and controversies. Embedded within society's scientific issues are ethical questions that need to be addressed and discussed in reasoned, thoughtful ways. The only way for our students to be prepared for such participation in societal discourse is to have practice in their school years, and what better place than the science classroom, where they can come to make sense out of scientific concepts in personally relevant, meaningful, interdisciplinary contexts.

Pragmatically, we as science educators do not have a choice but to move our curricula into a new era of contextualized science; our students want to have a say in local and global issues and with their unbridled access to information, they need to be able to identify the positions and motivations of various stakeholders. They also need to be well versed in analyzing the quality of information they receive, including the difference between science and pseudoscience. Today's students will find it necessary to understand the global implications of scientific decisions, not only in terms of the scientific repercussions, but also the economic, political, sociological, and ethical impacts. Finally, our students must possess the capacity to see beyond themselves and be able to relate to others' perspectives. This type of thinking requires opportunities for role-play and multiple forms of discourse. Such discourse, if explicitly taught and practiced, can guide students' thinking and decision making, and can be generalized to future, novel contexts. Moreover, practice in this type of knowledge generation models the type of discourse that real scientists practice on a daily basis, and precisely the type of reasoning essential for analyzing scientific problems.

Socioscientific issues–based curricula combine all of the above aspects in order to provide students with meaningful, relevant contexts for scientific learning. SSI is the use of ill-structured social issues with scientific connections to immerse students in real scientific argumentation and reasoning. While argumentation is not the only form of discourse in the SSI classroom (see SSI and Classroom Discourse section, p. 14), it is, nonetheless, a critical component. It is critical in (at least) two senses:

First, learning to argue well is intimately tied to learning to think well; second, the development of science and scientific understanding is predicated on the ability to convince, justify, and defend both through rhetoric and evidence-based reasoning. The importance of argumentation in science education has been well documented in the literature (Erduran and Jimenez-Aleixandre 2008; Osborne, Erduran, and Simon 2004).

To better understand what a curriculum based on SSI entails, it is helpful to contrast it with what it is not. Table P1.1 presents these contrasting features:

TABLE P1.1.

Features of an SSI Curriculum

Socioscientific Issues (SSI) Curriculum *is*	Socioscientific Issues (SSI) Curriculum *is not*
A research-based, interdisciplinary approach that enlists higher order problem-solving, argumentation, and research skills to analyze challenging, contextualized scientific concepts and issues.	A "cookbook" approach to scientific exploration that emphasizes "one right method" and predictable outcomes.
A method that uses real-world scenarios and real data in order to prepare students for their future roles as societal decision makers.	Simplistic use of hypothetical scenarios that are irrelevant to students' lives.
A conduit for scientific argumentation and discourse skills that mimic the manner in which real scientists research, discuss, debate, and deliberate scientific issues.	Emphasis on esoteric debates that allow students to contribute opinions rather than evidence.
A relevant and meaningful context for probing students' moral/ethical beliefs on controversial issues while guiding them to become tolerant and open to conflicting opinions and perspectives.	Reliance on "safe" subjects that avoid emotional connections and moral/ethical dilemmas.
A logical approach for modeling nature of science including the tentativeness of scientific conclusions, the importance of rational argument and skepticism, the role of creativity, and the distinction between science and pseudoscience.	A traditional approach to scientific methodology, which fails to recognize the varying social, contextual, and personal influences that contribute to scientific progress.

The distinctions presented in Table P1.1 firmly align the SSI curriculum with that of contemporary progressivism in science education. This view focuses on a student-centered environment that fosters responsibility and autonomous learning through actions and authentic learning experiences (Zeidler, Applebaum, and Sadler 2011). Let us examine several implications derived from this contemporary progressive view as it informs the craft of science teaching.

Distinction From Science, Technology, and Society

It is important to note that the SSI framework goes "above and beyond" past notions (at least how typically practiced) of science, technology, and society (STS) education. While STS education emphasizes the interrelationships among science, technology, and society, it seems to lack a theoretical framework that informs teachers and those involved in program development of pedagogical strategies that acknowledge the social and emotional development of children's identity as part and parcel with the curriculum. In our conceptualization, *socioscientific issues* is a broader term that subsumes all that STS had to offer, while also considering the ethical dimensions of science, the moral reasoning of the child, and the emotional development of the student" (Zeidler, Walker, Ackett, and Simmons 2002, 344). Therefore, our stance is that while STS does provide a context for connecting science and technology to society, SSI extends that context with clearly defined aims that foster a broader socio-cultural view of scientific literacy (Zeidler, Sadler, Simmons, and Howes 2005).

SSI and Scientific Literacy

Roberts (in press) discusses two views of scientific literacy. Vision I is concerned with promoting the academic content goals of scientific enterprise, while Vision II stresses broader goals related toward promoting a public understanding of science. The Vision II perspective is broader in scope and entails discourse about scientific issues that are contextualized in personally relevant ways. We position SSI as an extension of Vision II and we emphasize the notion of "functional scientific literacy." Our view affirms that our students cannot be fully "functional" when it comes to acting on matters of scientific importance without attention to the larger moral and ethical issues that surround those issues. So, our approach to developing functional scientific literacy through SSI is best understood as pragmatic endeavor, one that is more inclusive and addresses the psychological, social, and emotive growth of the child (Zeidler and Sadler 2011). We will find it includes the evaluation of moral and ethical factors in making judgments about both the validity and viability of situated scientific data and information relevant to the quality of public well-being and environmental stewardship of the biological and physical world.

Integrating Science Content

Our working assumption within the SSI framework is that SSI units of study afford the context for students to understand, through carefully crafted experiences, that scientific knowledge is theory-laden and socially and culturally constructed. The extent to which students internalize this depends, of course, on their developmental readiness. The process of experiencing science "in the making" would look different across varied grade levels. However, our central approach remains essentially the same regardless of grade level.

The teacher's role becomes secondary (but not less important) in relation to the SSI, which provides the social context for understanding scientific content, and the inquiry methods and reasoning skills students bring to bear on working their way through the issues. The teacher must learn to direct, prod, orchestrate, question, and facilitate, but it is clearly the students' engagement in the issue that is of central importance.

Cross-Curricular Connections

One of the advantages of an SSI curriculum, particularly at the elementary and middle school levels, is that it lends itself well to interdisciplinary connections. Many educators feel there is not enough time for science in elementary grades. However, a carefully designed SSI topic can involve a mix of reading skills, science content, social studies, mathematics, and art. As students get older, their education becomes increasingly focused and insulated, a process many believe reduces the overall effectiveness of science education. SSI units encourage the integration of scientific and nonscientific disciplines rather than their separation, which helps provide students with real, meaningful social contexts. That context, in turn, provides motivation to learn science content by making it seem more relevant and interesting.

SSI and Character

Moral education and its related forms of character education presuppose the formation of conscience (Zeidler and Sadler 2008). By this we mean that in the process of cultivating scientifically literate citizens, our aim is to foster the formation of a collective social conscience. The goal is to instill the desire to consistently hold one's actions up for internal scrutiny (i.e., reflective reasoning), which is a fundamental feature of conscience. It is no doubt a difficult task to not only grasp what is meant by "features" of conscience, but to aim instruction in a manner that nurtures it. Furthermore, it is also a given that no single definition of character can include *all* of the features that would appease philosophers and educators alike. However, we believe that Table P1.2 serves to capture at least a partial, pragmatic sense of the meaning of character, something achievable over time in the SSI classroom. In considering this table, our aim is to provide you with an understanding of some indicators that typically reflect what is, and is not meant by *character*.

TABLE P1.2.

Elements of Character

Character *is*	Character *is not*
An inclination to do what is right regardless of personal cost	Doing what others say is right to minimize personal cost
A will to respond to events according to values and principles	Responding to events based on urges, whims or impulses
A desire to seek habitual excellence in one's actions	Repeating mistakes in an uncaring manner
A commitment to engage in improving social order	Willfully ignoring social injustice
Actively hearing the words of others	Passively listening to the words of others
Displaying open-mindedness	Exhibiting close-mindedness
Normation: the conscious act of forming one's behaviors organically in terms of how they fit with society	Socialization: the passive act of accepting group behaviors mechanically with blind obedience
Acting thoughtfully (reflectively)	Acting without thought (unreflectively)
Responding to different situations in a flexible manner by understanding situation-specific demands	Responding to different situations in a fixed manner by not acknowledging situation-specific demands
Moral excellence: living up to one's full potential	Amoral or immoral mediocrity: falling short of one's potential

By participating in carefully designed, socially responsible activities, students will hopefully develop or have reinforced such qualities as reliability, trustworthiness, dependability, altruism, and compassion. SSI education requires contextualized argumentation, which provides an opportunity to practice education for citizenship. Democratic group decision making, facilitating understanding, fostering human values and caring, and nurturing emotional intelligence are central in an SSI classroom and recognized as building blocks of character (Berkowitz and Grych 2000; Wellington 2004).

Our recent research has shown that teaching within the context of socioscientific issues can increase students' moral sensitivity, thus contributing to overall moral development (Fowler, Zeidler, and Sadler 2009). Students have been shown to recognize and be concerned with the lives, health, and well-being of other people (Sadler 2004). Further research has confirmed that SSI approaches are able to develop students' sense of compassion toward diverse people in underdeveloped countries who are either alienated by the benefits of advanced technology or who are vulnerable to the dangers of its known or unintended effects (Lee et al. 2012). Within this work, we find that students' use of emotive reasoning, sympathy or empathy does not detract from the quality of making decisions about scientific issues. On the contrary, it enhances students' perceptions of ecological worldviews, socioscientific accountability and social and moral compassion. Hence, carefully crafted SSI programs can help develop students' socioscientific reasoning skills, including abilities such as recognition of the complexity of SSI, examine issues from multiple perspectives, and exhibit skepticism about information (Lee et al. 2012; Sadler 2011). These abilities are not trivial matters for they allow students to develop character and values that facilitate global conceptions of social justice.

SSI and the *Next Generation Science Standards* (*NGSS*)

In articulating their vision for K–12 education in natural sciences and engineering, the authors of *Frameworks for Science Education* (NRC 2012) cited the importance of preparing students to "engage in public discussions on science-related issues, to be critical consumers of scientific information related to their everyday lives, and to continue to learn about science throughout their lives" (pp. 1–2). This sentiment echoed the National Science Teachers Association's official position statement, "Teaching Science and Technology in the Context of Societal and Personal Issues" (2010) which advocates "that we not only know, understand, and value scientific and technological concepts, processes, and outcomes, but that we are able to use and apply science and technology in our personal and social lives." These documents emphasize functional scientific literacy (Zeidler and Sadler 2011), that is, the ability to recognize and use science in everyday contexts in order to ensure an educated workforce and informed citizenry.

The *Next Generation Science Standards* (*NGSS*) (NGSS Lead States 2013) emphasize the importance of functional scientific literacy in their opening statement:

> There is no doubt that science and, therefore, science education is central to the lives of all Americans. Never before has our world been so complex and science knowledge so critical to making sense of it all. When comprehending current events, choosing and using technology, or making informed decisions about one's healthcare, science understanding is key. (p. 1)

Clearly, the trend in contemporary science education is toward contextual, issue-based science—what Roberts (2007) refers to as Vision 2 science literacy—which moves away from inward-looking, canonic lists of what students should know about science to competency in science-related situations that students are likely to encounter in their everyday lives. But how do we foster this type of science literacy? While the *NGSS* do not advocate a particular curriculum, we believe that the SSI framework is uniquely situated for implementing the standards. In order to "make our case," let's examine the key elements of the *NGSS* and compare them to research on SSI.

The *NGSS* are built upon three major dimensions: scientific and engineering practices; crosscutting concepts that unify the study of science and engineering; and core ideas in the major disciplines of natural science (NGSS Lead States 2013, p. 1). In describing the *NGSS*, the authors highlight several "conceptual shifts" (Appendix A, p. 1) that distinguish the *NGSS* from the existing standards. One of these shifts states that "science education should reflect the interconnected nature of science as it is practiced and experienced in the real world" (Appendix A, p. 1). This mandate stems from criticism that current "cookbook" approaches to student investigations gives students a dull, simplistic perspective on the nature of scientific inquiry. It also reflects concerns that students are frequently unable to apply their science learning to practical situations. SSI curriculum addresses these critical concerns by providing students with real-world evidence about problems relevant to their lives (Zeidler and Keefer 2003). While examining and wrestling with controversial issues, students can increase their science content knowledge (Klosterman and Sadler 2010) and gain appreciation and understanding of the nature of scientific inquiry, including its complexity and interdisciplinary nature (Sadler, Chambers, and Zeidler 2004; Zeidler, Sadler, Applebaum, and Callahan 2009). Moreover, SSI curriculum acknowledges that questions about real-world issues related to science cannot be answered by science alone, but rather, require consideration of personal, ethical, and cultural factors as well (NGSS Lead States 2013, Appendix H). Through SSI, students are challenged to examine their core beliefs and are encouraged to, "approach decisions in an open unbiased way, respecting and acknowledging different perspectives, views, beliefs, and other ways of knowing" (NSTA 2010). This approach prepares students for science-related decision making in an increasingly pluralistic, global society and provides them with a greater understanding of the power, and limitations, of science.

Another conceptual shift in the *NGSS* is the organization of content around core ideas rather than specific facts within each content area. In this way, students can apply their knowledge to new information, like "experts," rather than by simply knowing isolated facts like "novices" (NRC 2012, p. 25). This approach fosters students' flexibility in problem solving, giving them the tools to apply their knowledge to novel situations. This is, of course, essential in an age when much of what students learn in science class in school becomes obsolete shortly thereafter. Focusing on core ideas in greater depth through extended inquiries also supports what we know about how children learn. Research on learning and cognition in science has suggested that science experiences should be "less episodic and fractured" (NRC 2001) than has been the norm and should extend over multiple classes allowing for deeper exploration of the subject. SSI curriculum requires students to explore topics in depth and in an integrated manner, a mechanism that helps students to understand the complexity of scientific issues and endeavors (Sadler, Barab, and Scott 2007) and reinforces both content and process skills. Arguably, any and all content standards for science could be addressed through careful selection of issues since, when using SSI, the science content remains the same as more traditional approaches, yet the context is transformed into something meaningful and emotionally provocative for the student so that content is better retained (Zohar and Nemet 2002).

One of the most striking shifts in the *NGSS* is that they present engineering concepts and practices at the same level of importance as science concepts and practices. The justification for this shift, which effectively "raises" engineering to the level of science (NGSS Lead States 2013), is that engineering and technology have a profound effect on civilization and students need to appreciate and analyze both the costs and benefits of various critical societal decisions (Appendix J, p. 4). Many SSI specifically deal with the trade-offs involved with technological advances and help students recognize different perspectives on the issues (Sadler, Barab, and Scott 2007). For example, SSI that deal with genetic engineering, space travel, or natural resource depletion all require analysis of the impacts of technology on society and society on technology. An SSI unit on these topics would also require students to examine the moral implications of technological advancements, raising the level of discussion beyond *can* we do something to *should* we do something. While SSI curriculum does not advocate any particular sides on issues, it helps students to recognize the complexity of issues in both scientific and human terms.

A most promising aspect of the *NGSS* is that they are aligned with the *Common Core State Standards (CCSS)* in English language arts and mathematics. This provides teachers with the opportunity to develop interdisciplinary lessons, work with colleagues from other subject areas, and of course, provide students with seamless learning and reinforcement of key concepts throughout their day. This movement away from siloing subjects supports research on how students learn and how content is mastered (Drake and Burns 2004) and also emphasizes the commonality of classroom discourse in all subjects. For example, the NGSS explicitly cite that common to the *NGSS*, *CCSS* for English language arts, and *CCSS* for mathematics is that, "students are expected to engage in argumentation from evidence; construct explanations; obtain, synthesize, evaluate, and communicate information" (Appendix D, p. 22). SSI is a powerful tool for helping students to develop these skills, particularly the ability to analyze evidence-based explanations (Walker and Zeidler 2007), a skill increasingly included in science assessments. Students are now expected to evaluate claims based on evidence, including contrasting and disconfirming evidence, differentiate between fact and opinion, and assess the credibility of the data. For example, the Programme for International Student Assessment (PISA), an international assessment program specifically designed to assess scientific literacy, requires students to "draw appropriate conclusions from evidence and information given to them," "evaluate claims made by others on the basis of the evidence put forward," and "distinguish personal opinion from evidence-based statements" (OECD 2009, p. 127). In their study, Sadler and Zeidler (2009) found strong overlap between the science literacy goals of the PISA and the strengths of SSI curriculum. They further argued that SSI provided more nuanced sociocultural and sociomoral experiences above and beyond the goals of PISA. Given the explicit mention of evidence-based argumentation in the *NGSS*, it is likely that it will be a focus of future assessments based on the *NGSS*.

The collaborative nature of SSI curriculum, which encourages communication between students through role-play and writing, seems tailor-made to, "prepare for and participate effectively in a range of conversations and collaborations with diverse partners, building on others' ideas and expressing their own clearly and persuasively" (NGAC and CCSSO 2010, p. 48). And while the NGSS are not explicitly aligned with the National Social Studies Standards (NSSS), overlaps do exist. For example, the *NSSS* include the following directive: "Social studies programs should include experiences that provide for the study of relationships among science, technology, and society" (NCSS 2010). This is an obvious strength of the SSI framework as SSI is based on the interplay among science, technology, and society, along with additional emphasis on moral education and the development of character.

Perhaps the most fundamental conceptual shift identified by the authors of the *NGSS* is that they are designed to "prepare students for college, career, and citizenship" (NGSS Lead States 2013, Appendix A). This statement reflects a bold effort to repair the leaky STEM pipeline in our nation by calling for science educators to simultaneously prepare students for global competitiveness, employability in STEM; and informed, responsible, participatory citizenship. While this is a tall order, it is achievable with curricula and pedagogy that flexibly support an array of skills and values. SSI does just that; its unique emphases on in-depth content mastery, evidence-based argumentation, nature of science perspectives, and moral development address multiple goals expressed in the *NGSS*. A summary of the key "Conceptual Shifts" (NGSS Lead States 2013, Appendix A) identified in the *NGSS* and the manner in which they are addressed by SSI is presented in Table P1.3.

TABLE P1.3.

How SSI Addresses "Conceptual Shifts" Identified in the *NGSS*

Conceptual Shift in *NGSS*	How SSI Addresses Conceptual Shift
"K–12 science education should reflect the interconnected nature of science as it is practiced and experienced in the real world."	SSI provides students with real-world evidence about problems relevant to their lives and helps students gain appreciation and understanding of the nature of scientific inquiry, including its complexity and interdisciplinary nature.
"The NGSS focus on deeper understanding of content as well as application of content."	SSI teaches content in context through extended inquiries that reinforce learning and encourage application of learning through numerous methods including lab and field investigations, internet research, debates, writing activities, and so on.
"Science and engineering are integrated in the NGSS."	Many SSI relate to engineering concepts and allow students to deliberate on the impacts of technology on society … and society on technology.
"The NGSS are designed to prepare students for college, career, and citizenship."	SSI prepares students to be critical consumers of scientific information and "rehearses" skills necessary for college and beyond including argumentation and discourse, evaluation and analysis of primary sources, understanding of diverse perspectives, and informed decision making.
"The NGSS and Common Core State Standards (English Language Arts and Mathematics) are aligned."	SSI is interdisciplinary in nature; it reinforces language, literacy, and math skills through evidence-based argumentation, research and writing, debate, data analysis … all in a manner that emulates real-world scientific application.

As the field of science education stands at the precipice of reform and revision, SSI curriculum provides a timely framework for addressing many teaching, performance, and content goals. More importantly, it prepares our students to be thoughtful, scientifically literate citizens who are prepared to be ardent participants in personal, local, and global decision-making.

The SSI Teacher and Classroom

At first blush, the decision to incorporate SSI into your practice using strategies that affect students in many positive ways seems like a potential panacea. At second blush, the task may seem daunting. We suggest that it is neither. However, experience has taught us that having an understanding of the conceptual framework that undergirds SSI provides a reasonable foundation to move forward toward sensible classroom implementation. The conceptual background above, coupled with more specific explanations and viewpoints of practice below, we believe, provide a solid justification to enact the SSI framework where natural connections to the subject matter you teach are found.

SSI and Pedagogy

Role of the Context (SSI Context)

Teachers looking to the web for SSI fodder may recognize that internet and issues-based learning activities can also be an invaluable resource in terms of exposing students to diverse perspectives on current scientific reports and claims. Students can spend their time reading and evaluating the multiple perspectives of a given socioscientific issue instead of "surfing" through a plethora of sometimes misleading information. Of course, this requires that teachers invest the time upfront to find both reliable as well as potentially unsound sources of scientific data and perspectives, so students may be confronted with mixed evidence and learn to assess the validity of varied claims and data.

Role of the Teacher

Perhaps one of the greatest advantages to SSI is that it allows the teacher to act as a facilitator to learning rather than simply a lecturer or conveyer of information. This frees the teacher to have more personal and frequent interactions with students and groups, and allows for ongoing assessment of student progress. A teacher engaged in SSI needs to rely on research and current information about a given topic to better direct classroom debates through thoughtful lines of questioning and probing. The importance of exposing students to discursive activities in the science classroom cannot be overstated if our goal is to ensure robust scientific literacy. Putting together an SSI module does not simply mean selecting a scenario where science or technology can "save the day." It does however mean two things: (1) You must practice and refine their craft, which requires skill and forethought, in order to align the SSI with course objectives; and (2) You need to be sensitive to the developmental and sociocultural needs of their students. Hence, to best enact the features of a SSI curriculum, as shown in Table P1.1, requires an understanding that you teach *children*, rather than content.

Role of the Students

Student impediments to success in the implementation of SSI tend to include moral (core) beliefs, scientific misconceptions, lack of personal experiences, weak understanding of content knowledge, underutilized scientific reasoning skills, and emotional maturity. In presenting this list, we do not mean to dissuade teachers from attempting an SSI approach. In fact, it is our position that insofar as students have such impediments, that we, as science educations, have a moral imperative to provide them with experiences that challenge their personal belief systems about the social and natural world. Doing so provide an opportunity for them to examine and evaluate their views, squaring their views with both scientific knowledge and other perspectives about those worlds. The ability to do so leads to more authentic and responsible learning.

SSI and Classroom Discourse

Sociomoral Discourse

Sociomoral discourse is a central necessity when issues of inquiry, discourse, argumentation, and decision making become a focal point in an SSI classroom. It occurs when one student's reasoning influences that of another, and, in return, a reciprocal relationship is forged. The dissonance compels students to negotiate, resolve conflicts, and enhances the quality of their own arguments. Such discussions have been described in the literature (e.g., Berkowitz 1997; Zeidler and Keefer 2003) and have proven to enhance the quality of reasoning by providing varied viewpoints that require the use of counter-positions, evidence, and just solutions over the course of development.

Argumentation and Debate

The inclusion of argumentation and debate in the science classroom is a rising area of interest among science educators just as issues of social controversy in science are proliferating with the advancements of technology. Using argumentation and debate, however, is a useful means to engage thinking and reasoning processes, and to mirror the discourse practices used in real life in the advancement of intellectual and scientific knowledge. For the purposes of the classroom practice, a focus on tolerance, mutual respect, and sensitivity must be modeled and expected.

Discussion

Productive debate and argumentation is not always practical or even possible in every educational setting, particularly for educators with little experience managing it. Teachers may first consider guided discussions rather than debate. Such discussions can allow educators to address controversial socioscientific topics in a more controlled manner, which may be especially helpful in certain contexts.

Critical Thinking

Whether business or politics or both motivate concerned citizens, calls for increased scientific literacy typically include a plea for the education system to produce students who are critical thinkers. One of the benefits of including an SSI curriculum is that the discussion and debate of controversial socioscientific issues necessitates that students develop many of the skills and dispositions associated with critical thinking. The core creative-thinking skills of analysis, inference, explanation, evaluation, interpretation, and self-regulation (Facione 2007) will all be encouraged by SSI units as will the dispositions associated with them. Incorporating SSI can therefore help to produce students who are evidence-seeking, open-minded, analytical, systematic, judicious, and increasingly confident in their reasoning.

Summary (Take-Home Message)

For preservice and practicing teachers, the realization that science education for many (most) students has included years of indoctrination, dogmatism, or authoritarianism is a sobering epiphany. However, there is no place in science, and therefore no place in science education, for the protection of concepts and theories from criticism. The challenge for science teachers is to allow students to have personal experiences that do not immediately negate their belief systems; rather, the aim is to provide the conditions necessary to enable the development of a personal epistemology through continued exposure to, and interaction with, the nature of science and SSI.

The use of argumentation and relevant SSI as a framework for science class curricula is essential for enabling scientific concepts to enter students' individual belief systems. The fatal flaw held by many teachers is their own pedagogical belief that concepts can be taught using sufficient explanations and tidy analogies that will then magically alter students' core beliefs. The use of SSI strategies challenges students to reevaluate their prior understandings, providing an opportunity for them to restructure their conceptual understanding of subject matter through personal experiences and social discourse.

Voices From the Field: What Practicing Teachers Can Tell Us About SSI Classrooms

Not surprisingly, the SSI classroom will likely be a place of lively discourse, collaborative inquiry, and authentic learning. Crucial to the successful implementation of this approach is the recognition that the teachers' role becomes that of a catalyst, facilitator, antagonist, and mentor. To fulfill these roles is to transform one's teaching practice to be consistent with that of a student-centered classroom. The SSI teacher needs to be vigilant, challenging students' core beliefs about their understanding of how scientific evidence is used in persuasive arguments. The teacher must be willing to model good rhetorical skills, maintain mutual respect in the face of differering interpretations and opinions, demonstrate the need to scrutinized assumptions about research, question authority, and maintain a healthy skepticism.

We have followed several strategies (King and Kitchener 2004; Zeidler, Applebaum, and Sadler 2011) that have provided useful guidelines in maintaining a safe student-centered environment. These include:

1. Modeling respect for students' assumptions about knowledge, thereby providing emotional support for them to take intellectual risks.

2. Be willing to discuss controversial issues, make sure they connect to other areas of subject matter, and make available to students resources that report empirical claims and lines of reasoning from multiple perspectives.

3. Provide multiple opportunities for students to examine and analyze evidence, as well as each other's viewpoints, in order to discuss, defend, and possibly reform their viewpoints.

4. Teach students strategies for systematically gathering evidence (forms of data), assessing the relevance of the data, evaluating data sources, and forming judgments or positions informed by that data.
5. Help students to make connections to the Nature of Science in understanding how science is empirically based, culturally influenced, developmental and creative, and how knowledge may be tentative or subject to change with the introduction of new data and ideas.

Typically, teachers who consider an SSI approach are prompted to do so because they are either dissatisfied with the inherent limitations of conventional classroom learning, or simply wish to build upon and extend past successes. Such teachers do so because after examining the case and evidence for SSI, they see the benefits such an approach can bring to their students. However, as with any new pedagogical strategy, there are, no doubt, some challenges that must be confronted. Concerns might also highlight issues related to student assessment. The controversial nature of SSI might also give pause in terms of possible objections from parents or administrators. And how does one get started implementing SSI? These are all legitimate and significant issues to consider. We thought that our thoughts on these matters would have more credibility if coupled with voices of practicing teachers who also had to wrestle with the same issues. It is enlightening to hear what their voices have to say.

We asked several teachers, each committed educators who have made the decision to implement SSI into their science curriculum, to respond to what we believe are six important questions related to both that decision and corresponding pedagogical issues. We think it is instructive to hear the voices of practicing teachers who are at the forefront of using SSI-based instruction. We asked these particular teachers if they would like to share their experiences for the benefit of other science teachers who are considering using the SSI approach. We also thought it would be interesting to present the thoughts of teachers from two different cultural perspectives—both in the United States and South Korea. Some have more experience than others in using SSI, but their perceptions are all relevant. To highlight the natural progression of SSI perspectives our teachers convey, we begin with Tom Dolan, who teaches at the elementary level, followed by Hyunsook Chang and Kisoon Lee, who teach at the middle school level, and lastly Scott Applebaum, who teaches at the high school level. After the teachers' responses to each question are presented, we offer our own interpretation and summary of those responses.

(1) What *prompted* you to begin using SSI in your classroom?

Tom: Throughout my teaching career, I have been mindful of Isaac Newton's third law of motion, which states, "For every action, there is an equal (in force) and opposite (in direction) reaction." This concept is fascinating to me – the thought that single, isolated forces cannot happen. The everyday forces that surround us always occur in pairs, they are always equal in magnitude and they are always equal in time. In the academic world, this law applies most often to the scientific happenings encountered in daily life, though for me, there is a strong, personal, philosophical relevance as well. I believe that the greater the teaching force I can project that extends beyond the boundaries of my classroom, the greater the force of student learning. That is where SSI comes into play for me.

For me, SSI offered a new (and often uncharted) avenue in my approach to elementary science education. There is a certain presumption associated with thinking that a science teacher alone can provide all the necessary science information that students need, while simultaneously teaching them to think critically and helping them to become lifelong learners. The reality is that educating—truly educating—involves the pursuit of more cooperative curriculums that allow for the inclusion of both independent and collective research, and the engagement of and partnership with scientific professionals, parents and community members. In my mind, SSI offers that type of forum. It moves beyond the plethora of "new" scientific teaching methods that are muddled with past approaches and simply amount to the "same old stuff" packaged in new branding. It truly is a curriculum approach that uniquely balances the acquisition of scientific knowledge and the development of moral reasoning.

Hyunsook: When I first started teaching, I joined a (Korean) science teachers' association at the invitation of my professor. That was when I first encountered SSI. At first, I thought that SSI was just another teaching method and materials. But as I continued to participate in the association, I came to learn that SSI is not just about teaching materials, but could be a critical tool to nurture qualified citizens in society. Of course, as members of the society, we need to be open-minded and have broader views to be more receptive to various opinions. Instead of just arguing against each other, if we try to understand broader perspectives, we can take a step closer to resolving problems.

Kisoon: I was searching for science teachers' association to find ways to teach science in a more professional way. Once participating in a science teachers' association, I found teachers there were interested in the issue of 'democratization of science' and they made various efforts to actually deal with 'socio-ethical aspect of science' in science education. During this time, I came to think I wanted to use SSI for three reasons:

1. Students were very satisfied with and actively participated in the class using SSI. Students said: "This is exactly the science class that we need to have." And seeing them join the class so actively and have fun in there I felt really great as a science teacher. Different from conventional class, SSI class is more open for students to express their opinions and closely relates to our daily life. I believe that is why students are more actively participating in the class.

2. I myself learned a lot about the socio-ethical aspects of science. I didn't have much interest in social issues but with SSI class, I was able to learn more about socio-ethical aspects of science and I was so happy with that.

3. The way students discuss and reach an agreement in SSI was quite similar to the process actually happening in the society. Studying how the discussion progresses and develops, I became sure that this can contribute to deepening our understanding of the way people react to and handle socio-ethical issues of science. These are the reasons why I keep going with SSI class.

Scott: My primary goal, as a graduate student at the University of South Florida School of Education, was to discover teaching skills to become an excellent secondary science educator. After 25 years of practicing dentistry, I discovered that my return to high school as a teacher would be an effortless transition because science class instruction methods hadn't changed. After one year of teaching, I learned that the students had great capacity to learn intricate science concepts; however, they were unable to correlate the information with contemporary topics and did not retain the knowledge for more than a few weeks. It occurred to me that these same individuals could learn complex computer technology without written or formal instruction, as well as the lyrics of a thousand songs. Perhaps the context and delivery of information in the classroom was flawed. The following year, I was invited to attend the NARST conference in Vancouver, where several professors and graduate students presented proposals for a new pedagogy that aligned with 21st-century science and provided an interesting format of delivery of science concepts to high school students.

Our thoughts: It is interesting to note that for the most part, these teachers were genuinely dissatisfied with the current state of more traditional teaching practices that have not changed very much for many years. They tended to have a predisposition to bend toward the acceptance of a more sociocultural approach to contextualizing science in a democratic manner. They were exposed to SSI through professional associations or working with science educators and have come to see that this type of an approach positions both the students as well as the teacher in the act of learning.

(2) What do you see as the *benefits* of SSI?

Tom: The benefits of incorporating SSI into elementary science curriculum are threefold in that SSI has the potential to increase critical-thinking skills, allow students to analyze moral dilemmas associated with real-world social issues, and enhance overall student learning through embedded character education. Through its use of use of controversial topics, SSI promotes self-evaluation, reflection, and mutual respect of differing points of view.

Hyunsook: SSI lets us know that the social issues in the society are not something we can easily resolve. There are various opinions and different points of view and every aspect has its own reasons and basis for the argument. The deeper and wider it goes for a certain issue, the more complicated and the broader it goes to touch ethical and moral aspects; we cannot take a simple one-sided approach. This seems to be an important lesson to help us getting better in living our daily lives.

Kisoon: SSI is meaningful in that students can develop the capabilities and understanding to be a citizen or a leader in the science and technology society of the future. It can also be an opportunity to learn discussion skills and create knowledge as [these are] the main teaching tool in SSI class.

Scott: The mission statement of most school districts includes a goal of encouraging students to become "lifelong learners." Because learning new information during the majority of an individual's life is not through note taking from formal lectures, PowerPoint presentations, and 20 chapter textbooks, perhaps the instruction in the classroom should replicate the manner in which information is presented outside the classroom. SSI instruction encourages students to learn complex science concepts, by presenting the knowledge in a context that is personally relevant, socially shared, contentious, interesting, and contemporary.

Our thoughts: These teachers share the view that students need to develop lifelong learning skills. The exercise of critical thinking and self-reflection, promoted in SSI approaches, allows students to practice the kinds of reasoning that will enable them to better navigate uncertain waters. References to the development of positive character traits, as in the ability to co-construct knowledge and the importance of understanding multiple viewpoints, for example, are clearly emphasized by these teachers. The teachers are aware that nurturing these abilities can help encourage students to assume future citizenship responsibilities.

(3) Are there any particular *challenges* in implementing SSI? If so, how do you address them?

Tom: I think perhaps the biggest challenge to implementing SSI is maintaining the goal of teaching the children "how" and not "what" to think. As adults, our worldview is somewhat solidified, in that our perceptions, experiences, and education have helped us make some sense of our complex realities. Our behaviors and values are developed in ways that elementary students are not. As such, it would be easy to align SSI exercises with our core beliefs and facilitate SSI reasoning in such a way that it reflects and subtlety imparts our worldview as "correct." But to do so would subvert the true value of SSI. The fact remains that the structure of SSI requires that moral dilemmas be examined from multiple perspectives so that scientific information can be properly analyzed and synthesized and social development can be strengthened. Holding steadfast to this perspective helps to ensure that educators refrain from becoming impediments to proper SSI implementation.

Hyunsook: Implementing SSI in class is not easy. As a teacher I have to follow the curriculum and it takes a lot to introduce something outside of the curriculum. So implementing SSI requires a serious decision made by myself. For middle school students, it took a lot to get them to think deeper on a certain issue.

Kisoon: In the class I asked students to find ways to iron out the differences and reach an agreement. They got disappointed when their opinion was not adopted and I saw sometimes that they just agreed to accept unwillingly. So whenever I closed the class, I reiterated the meaning of the process of reaching an agreement. "...In making decisions over socio-ethical issues of science, sometimes your opinion may not be adopted. But going through the process to reach an agreement can make a difference in the society. While discussing the issue to reach an agreement, one side may point out the possible side effects in the other's opinion and try to find out ways to avoid that. With these efforts, we could reach a more reasonable conclusion, just as we did in our class. If your opinion is adopted, you should accept the other's opinion as far as possible to minimize any social disparity, and if your opinion is not adopted, you have to accept the other's [but you can] serve as watch dog to make sure that it is actually working."

Scott: The challenge of SSI instruction includes the necessity of teachers having extensive knowledge of the topics, including information that is personally relevant, contemporary, contentious, and interesting. Students respond to SSI instruction when the teacher has commanding knowledge of the subject matter. The largest challenge is trusting that students can learn the information without formal instruction. SSI instruction needs to be presented in a student-centered classroom, with the teacher acting as moderator, guide, and referee. Students develop a sense of academic motivation when surrounded by classmates and teachers who are motivated. The comfort and security of textbook knowledge [only] has to be sacrificed to model authentic learning.

Our thoughts: Implementing a new and comprehensive strategy is never an easy task. SSI- based instruction challenges the normative roles of traditional teaching expectations; learning to "unlearn" well-formed teaching habits and assume a new role as facilitator of learning can be challenging. It is also difficult not to impose you own positions on the class. Teachers voice concerns about their ability to adhere to curricula standards, so approaching an SSI approach takes considerable deliberation and forethought about the role of authentic learning in the science classroom.

(4) SSI is very different from traditional text-based teaching. (a) Can you still prepare your students adequately for more standardized or traditional-based tests? Please explain. (b) Have you found other forms of assessment to be useful within the SSI framework? Please explain.

Tom: (a) The inclusion of SSI into the classroom does not encumber the preparation required for standardized and traditional-based tests. Because SSI involves employing the fundamentals of sociomoral discourse, argumentation, debate, and discussion, it is imperative that students have a solid comprehension of the elementary science concepts being discussed prior to implementing SSI techniques. Those elementary science concepts are the very same concepts contained on standardized and traditional-based tests. To that end, I normally incorporate a hands-on activity into each lesson that addresses each scientific concept prior to initiating any SSI exercise. Also, for what it's worth (putting aside my personal views regarding the value of standardized assessment), year-after-year, my students tend to fall into the top of the county in terms of test performance.

(b) I use a preliminary document to establish baseline data with regard to their ethical stances and personal opinions, and preconceived notions about a given scientific document (or concept). Once the SSI activity is completed, I then follow up with both a reflective opinion document and content examination document in an effort to determine content retention and the overall effect of the activity on subsequent moral reasoning. Because the answers provided on the preliminary and reflective documents are subjective in nature, I evaluate them using current moral development theory. I normally look at content knowledge in terms of content retention and argumentation development. For that, I use some sort of preset rubric. This type of analysis, though time-consuming, allows me to properly gauge how SSI is working in my classroom at any given time.

Hyunsook: (a) Since we have to take the test set by the national curriculum, of course I care about that as a teacher. I always try to teach what is required by the national curriculum and make more efforts to get my students to be ready for the traditional test.

(b) It was not easy to make assessment after implementing SSI. In most cases, it was an essay or report to see how students explain their position and outline the basis to back up the argument. But this tends to be an assessment on how knowledgeable they are about (the process of) SSI so it is not what I prefer. Using SSI in class is more to give students opportunities to think deeper and wider and not to neglect ethical aspects of an issue. Text-based assessment is not the right tool to assess this. Since it is challenging in many ways, I do not want to have assessment if I implement SSI class in the future. Having the class itself is meaningful enough, I believe.

Kisoon: In text-based assessment, it is difficult to have SSI assessment as the curriculum is mostly about concepts and how well students understand the concepts—and the assessment also should be centered on that. (In Korea the curriculum set by the government is the key to decide the scope of education in the field.) And since SSI values each and every opinion of individuals, it is challenging to have text-based assessment.

In activity-based assessment, it is possible to try SSI assessment. Since SSI is focused on various students' activities (role playing, discussion, debate, others) I gave scores depending on how actively they participated.

Scott: The obsession with standardized written tests and other forms of assessment has become a distraction to authentic teaching and learning. While school boards and administrators debate which science lesson plans make the best medicine, the students have been slipping into a classroom coma. When formal (classroom) education ends, these assessments cease to exist; however, learning continues to encompass the daily cerebral activity of "lifelong students." Countries, states, school districts, schools, teachers, and students are all subjected to grading, based upon scores derived from written tests. And yet, navigating successfully through life involves critical-thinking skills and not the retrieval of memorized facts. SSI instruction should include assessments in a form that evaluates student ability to apply scientific concepts to contemporary issues: measuring/evaluating discussion quality, discourse development, determining if students can document reliable investigation evidence, and developing activities that replicate authentic presentation of scientific issues. The first assessment is evaluating whether students recognize certain scientific information as being contentious in nature. Students can demonstrate learning new knowledge from negotiating SSI by creating posters that present both "sides" of an argument, developing discourse and debate speaking points from reliable evidence, and writing investigative essays that demonstrate their ability to recognize reliable information and successfully organize the information into a logical format. Through this process, their ability to remember esoteric information (on standardized and traditional tests) increases because the information makes sense to them.

Our thoughts: The teachers share a common priority: Students do need to perform well on centralized assessments that are district or state driven. While their implementation of SSI no doubt varies, they have confidence in the abilities of students to comprehend, retain, and apply scientific knowledge discovered through SSI pedagogy. The development of SSI-specific rubrics, while challenging and requiring more time and effort, provide authentic documentation of student growth and performance. The use of alternative forms of assessment also provides feedback to teachers who can then fine-tune their lessons and ensure proper content coverage.

(5) Does the *controversial* nature of SSI pose any problems with parents or administrators? If so, how have you been able to address them? If not, what has been the reaction?

Tom: I think, on paper, SSI at the elementary level can seem daunting to parents or administrators who are not familiar with its implementation. Parents can sometimes seem initially apprehensive when they learn about SSI because to them it appears to be some sort of indoctrination technique where children will be coerced into adopting particular worldviews. I do everything I can to assuage their concerns and I try to actively involve parents and administrators in the SSI learning process. In my experience, once they understand the true premise of SSI through both observation and participation, both parents and administrators have been supportive of my desire to include it in my curriculum endeavors. I tend to receive emails from parents indicating that SSI helps make science more interesting to their children, and that they are better prepared than other students for science curriculum imparted in future grade levels. Several days ago, I received the following e-mail from a parent of two of my former students that read, " Just thought I'd let you know that Jesse (now a ninth-grade student in the International Baccalaureate Program) used her SSI skills today in her debate and silenced the whole room." For me, there is no bigger compliment!

Hyunsook: I haven't experienced these kinds of problems related to the controversial nature of SSI, but sometimes I do feel cautious dealing with (certain) controversial issues. When we discuss nuclear energy, since Korea is still highly dependent on that energy and makes positive [commentary] to promote that, I feel cautious to express points of views against it. In one class, one of the students' father worked for the Korea Electric Power Corporation and the student asked his father about the discussion made during the class and later argued back after listening to his father making positive arguments about nuclear energy. Likewise, there was such strong support for the Stem cell research of Dr. Whang at the early stage, it was not easy to discuss [other aspects] since there was such strong support for the research because this was an area where Korea could lead in science—I felt cautious to present positions against it. (Note: Dr. Whang is a Korean scientist considered to be a pioneer, who conducted and published research on stem cells in high-profile journals, but later became discredited for faking some of his data and violating other ethical rules of research and civil conduct.)

Kisoon: I didn't experience any of those problems.

Scott: I believe many administrators would be resistant to change and certainly afraid that students would not be able to pass standardized end-of-year tests, which would reflect poorly on the school and its leadership. Disclosure to the parents of the proposed teaching method and goals should be made at the open house, after the second week of school. When my students bring home trivial and important information regarding the contents of sunscreen and shampoo and the threat of imminent death from consuming fried foods and mayonnaise, parents understand that their children are learning the practical application of scientific concepts. While parents have enthusiastically embraced SSI instruction, administrators have been less than accepting. On two occasions each school year, school administrators sit at the back of the classroom and evaluate the quality of classroom instruction. In this regard, the reviewer completes a predesigned form that was created by state and district administrators. The expectations provide that all science teachers must comply with specific guidelines of teaching methods and information distribution. Because the burden of providing factual knowledge is placed entirely upon the teacher, student performance and individual motivation is never determined and the goal of creating "learners" is lost. There are no score sheets that measure the quality of student-centered classroom learning.

Our thoughts: For this question, only one has not experienced any challenges related to the implementation of controversial SSI. These teachers are generally aware that presenting issues that challenge prevailing normative beliefs will have to be handled in a thoughtful manner. It seems clear to us that providing a carefully thought-out rationale to parents and administrations that are likely uninformed as to the benefits SSI offers students is a prudent strategy. As a professional, one should always be able to justify his or her choice of pedagogical means to achieve particular educative ends. In doing so, parents can become very supportive because they see first-hand the renewed enthusiasm their sons and daughters have toward science.

(6) How would you suggest a novice SSI practitioner get started? Any good tips?

Tom: Here is my list of tips:

(a) Before getting started, I would read a significant amount of research pertaining to SSI. There are various strategies for SSI implementation, and I would start with the one that feels most comfortable to the instructor.

(b) I place quite a bit of value on using a hands-on activity as a way to introduce the scientific concepts prior to SSI implementation. For primary students, this is particularly helpful in providing them with a solid content base for which to approach moral dilemmas.

(c) Relevance (to both students and course content goals) is key to successful SSI facilitation. Without both, SSI is much harder to implement.

(d) It is imperative that you hold students accountable for accuracy and for constructive, rather than deconstructive-based, argumentation.

(e) The beauty of SSI lies with the fact that it is open-ended in nature, and therefore, don't be afraid to allow the discussion to veer off into additional areas of scientific study.

(f) Continue to tweak your approach until you see the type of results you desire with regard to moral reasoning, social development, and scientific understanding.

(g) Have fun!

Hyunsook: It depends on what the motivation is for the teacher. Is he/she interested in social issues or is it simple to introduce and use (different) materials?

Kisoon: I had to go through lot of trial and error since I didn't have clear understanding of how to roll out discussions and debates. Get familiar with practical discussion skills through training; it will make your class a lot more substantial and satisfactory.

Adding to that, sometimes when you have SSI class, you run into a situation where the teacher's view is conflicting with the students' views. As a teacher you may want to persuade or try to convince your students (of your opinion), but instead I believe it is better to include this just as part of the class and give students right to choose.

Scott: Add an SSI activity to every section of instruction. Do not be afraid that students will not learn complex scientific principles unless it is presented in a lecture and a textbook. This method is most effective if the teacher believes the pedagogy is effective. Model the learning method by becoming directly and visibly part of the early investigations. Experience will reduce the discomfort and awkwardness … be patient!

Our thoughts: The unifying theme here is practice and patience pays off! But there is also an underlying theme, a subtext if you will, that is present. Interestingly, it is something that certain practitioners of karate also say: "Practice does not make perfect. Perfect practice makes perfect!" The meaning here is clear: Teachers must be willing to reflect on their practice and constantly strive to build and improve upon it. We think that the tips Tom offers provide very clear and practical guidelines for those new to SSI. We also hope that all of our teachers' voices resonate with your own inner voices of educational philosophy and pedagogy concerning the importance of sociocultural perspectives of science education.

Note:

Tom Dolan teaches at Pride Elementary School in Tampa, Florida.

Hyunsook Chang teaches at Kuksabong Middle School in Seoul, South Korea.

Kisoon Lee teaches at Sinam Middle School in Seoul, South Korea.

Scott Applebaum teaches at Palm Harbor University High School, in Palm Harbor, Florida.

We wish to thank these SSI teachers for their valuable professional insights and contributions to our book.

References

Berkowitz, M. W. 1997. The complete moral person: Anatomy and formation. In *Moral issues in psychology: Personalist contributions to selected problems*, ed. J. M. DuBois, 11–41. Lanham, MD: University Press of America.

Berkowitz, M. W., and J. H. Grych. 2000. Early character development and education. *Early Education and Development* 11 (1): 55–72.

Carnegie Corporation of New York. 2009. The opportunity equation: Transforming mathematics and science education for citizenship and the global economy. *http://opportunityequation.org/report*

Drake, S. M., and R. C. Burns. 2004. *Meeting standards through integrated curriculum.* Alexandria, VA: Association for Supervision and Curriculum Development.

Erduran, S., and M. P. Jiménez-Aleixandre, eds. 2008. *Argumentation in science education: Perspectives from classroom-based research.* Vol. 35. Netherlands: Springer.

Facione, P. A. 2007. *Critical thinking: What it is and why it counts*. Millbrae, CA: California Academic Press.

Fowler, S. R., D. L. Zeidler, and T. D. Sadler. 2009. Moral sensitivity in the context of socioscientific issues in high school science students. *International Journal of Science Education* 31 (2): 279–296.

King, P. M., and K. S. Kitchener. 2004. Reflective judgment: Theory and research on the development of epistemic assumptions through adulthood. *Educational Psychologist* 39 (1): 5–18.

Klosterman, M., and T. D. Sadler. 2010. Multi-level assessment of scientific content knowledge gains associated with socioscientific issues-based instruction. *International Journal of Science Education* 32: 1017–1043.

Lee, H., H. Chang, K. Choi, S. W. Kim, and D. L. Zeidler. 2012. Developing character and values for global citizens: Analysis of pre-service science teachers' moral reasoning on socioscientific issues. *International Journal of Science Education*. 34 (6): 925–953.

National Council for Social Studies (NCSS). 2010. *National curriculum standards for social studies. www.socialstudies.org/standards/strands*

National Governors Association Center for Best Practices and Council of Chief State School Officers (NGAC and CCSSO). 2010. *Common core state standards.* Washington, DC: NGAC and CCSSO.

NGSS Lead States. 2013. *Next Generation Science Standards: For states, by states.* Washington, DC: National Academies Press.

National Research Council (NRC). 1996. *National science education standards*. Washington, DC: National Academies Press.

National Research Council (NRC). 2001. *Classroom assessment and the National Science Education Standards*. Washington, DC: National Academies Press.

National Research Council (NRC). 2012. *A framework for K–12 science education: Practices, crosscutting concepts, and core ideas.* Washington, DC: National Academies Press.

National Science Teachers Association (NSTA). 2010. *Position Statement on Teaching Science and Technology in the Context of Societal and Personal Issues. www.nsta.org/about/positions/societalpersonalissues.aspx*

Organisation for Economic Co-operation and Development (OECD). 2009. PISA 2009 assessment framework: Key competencies in rreading, mathematics and science. *www.oecd.org/dataoecd/11/40/44455820.pdf*

Osborne, J. F., S. Erduran, and S. Simon. 2004. Enhancing the quality of argument in school science. *Journal of Research in Science Teaching* 41 (10): 994–1020.

Roberts, D. A. 2007. Scientific literacy/science literacy. In *Handbook of research on science education*, ed. S. K. Abell and N. G. Lederman. Mahwah, NJ: Routledge,Taylor, and Francis.

Sadler, T. D. 2004. Informal reasoning regarding socioscientific issues: A critical review of research. *Journal of Research in Science Teaching* 41 (5): 513–536.

Sadler, T. D., ed. 2011. *Socio-scientific issues in the classroom: teaching, learning and research* (Vol. 39). Netherlands: Springer.

Sadler, T. D., S. A. Barab, and B. Scott. 2007. What do students gain by engaging in socioscientific inquiry? *Research in Science Education* 37: 371–391.

Sadler, T. D., F. W. Chambers, and D. L. Zeidler. 2004. Student conceptualisations of the nature of science in response to a socioscientific issue. *International Journal of Science Education* 26: 387–409.

Sadler, T. D., and D. L. Zeidler. 2009. Scientific literacy, PISA, and socioscientific discourse: Assessment for progressive aims of science education. *Journal of Research in Science Teaching* 46 (8): 909–921.

Walker, K. A., and D. L. Zeidler. 2007. Promoting discourse about socio-scientific issues through scaffolded inquiry. *International Journal of Science Education* 29: 1387–1410.

Wellington, J. 2004. Ethics and citizenship in science education: Now is the time to jump off the fence: Ethics in science education. *School Science Review* 86 (315): 33–38.

Zeidler, D. L., S. M. Applebaum, and T. D. Sadler. 2011. Enacting a socioscientific issues classroom: Transformative transformations. In *Socio-scientific issues in the classroom,* ed. T. D. Sadler, 277–305. Netherlands: Springer.

Zeidler, D. L., and M. Keefer. 2003. The role of moral reasoning and the status of socio-scientific issues in science education. In *The role of moral reasoning on socio-scientific issues and discourse in science education*, ed. D. L. Zeidler, 7–38. Dordrecht, The Netherlands: Kluwer Academic Publishers.

Zeidler, D. L., and T. D. Sadler. 2008. Social and ethical issues in science education: A prelude to action. *Science & Education* 17 (8): 799–803.

Zeidler, D. L., and D. L. Sadler. 2011. An inclusive view of scientific literacy: Core issues and future directions of socioscientific reasoning. In *Promoting scientific literacy: Science education research in transaction,* ed. C. Linder, L. Ostman, D. A. Roberts, P. Wickman, G. Erickson, and A. MacKinnon, 176–192. New York: Routledge, Taylor, & Francis.

Zeidler, D. L., T. D. Sadler, M. L. Simmons, and E. V. Howes. 2005. Beyond STS: A research-based framework for socioscientific issues education. *Science Education* 89 (3): 357–377.

Zeidler, D. L., T. D. Sadler, S. Applebaum, S., and B. E. Callahan. 2009. Advancing reflective judgment through socioscientific issues. *Journal of Research in Science Teaching* 46 (1): 74–101.

Zeidler, D. L., K. A. Walker, W. A. Ackett, and M. L. Simmons. 2002. Tangled up in views: Beliefs in the nature of science and responses to socioscientific dilemmas. *Science Education* 86 (3): 343–367.

Zohar, A., and F. Nemet. 2002. Fostering students' knowledge and argumentation skills through dilemmas in human genetics. *Journal of Research in Science Teaching* 39: 35–62.

Part 2

Implementing SSI in the K–12 Classroom

Developing Your Own SSI Curriculum

Constructing an SSI unit may seem daunting at first, but it will no doubt become second nature with practice and experimentation. We suggest that teachers start by adding SSI lessons to already-existing units. For example, if you are doing a unit on the human body, consider incorporating research on a related controversial issue such as, "Should fluoride be added to drinking water?" or "Should fried foods be banned?" Taking small steps to incorporate controversial issues in your classroom will lay the groundwork for more expansive use of SSI in the future.

The SSI framework is flexible and allows for tremendous creativity and personalization to meet the needs, interests, and goals of teachers and students. To help facilitate this process, we have identified some steps that are useful in guiding teachers through the development and implementation of SSI in their classrooms:

1. **Identify Topics.** Review newspapers, books, internet sources, professional science education–related journals and television/movies for current issues related to your subject matter and course objectives. There are local and global controversies related to almost any science topic. As you explore

topics, consider your students' interests and select topics with relevance to their lives and the curriculum. Some possible SSI include:

Genetically Modified Foods	Cell Phones and Health	Sunscreen: Help or Harm?
Milk: Animal vs. Soy?	Medical Marijuana	Fast Food Limits
Animal Dissection in the Classroom	Alternative vs. Fossil Fuels	Paper or Plastic Bags?
Is Coffee Good for You?	Bicycle Helmet Laws	Stem Cell Therapy
Sex-Change Surgery	Fur Ban	Mandatory Fat Camps
Offshore Oil Drilling	Animal Research	Steroids in Sports
Alcohol Consumption	Hunting for Population Control	Beach Enrichment
Farm-Raised vs. Wild-Caught Salmon	Plastics and the Environment	Tap vs. Bottled Water
"Designer" Babies	Space Settlements	Global Warming
Deforestation	Vaccinations	Fluoride in Water
Reproductive Issues	Long-Line Fishing	Herbal Remedies
Antibiotics	Locating Landfills	Satellite Tracking and Privacy
Cochlear Implants	Smoking Bans	Fracking
Euthanasia	Exotic Animals as Pets	Land Use
Cloning	CFL vs. Incandescent bulbs	Texting and Distraction

2. **Collect Resources.** Look for a range of sources reflecting a diversity of viewpoints. Remember that many controversial issues have more than two sides, so it is essential to seek out resources that reflect the concerns of a variety of stakeholders. Provide resources across the bias spectrum, so that students have an opportunity to analyze the objectivity of the material.

3. **Introduce Topic.** Engage students with magazine headlines, articles, advertisements, YouTube videos, photos, models, or other media. Using a broad range of materials ensures that most students will be "hooked" into the topic and ready to engage in learning.

4. **Prepare Students for Discussions.** Set ground rules for class discussions, emphasizing the value of all ideas, mutual respect for participants, and intolerance for mockery or personal attacks. Collaboratively constructing discussion rules with students can be particularly powerful in developing

a classroom culture based on trust and mutual respect … a setting that will yield tremendous benefits during SSI discussions. Emphasize the importance of using evidence to back up opinions during discussions. Encourage students to examine the sources of their personal beliefs.

5. **Pose Controversial Questions.** Introduce contentious questions (e.g., "Should schools charge a fat tax for unhealthy foods?") and challenge "common knowledge" of subject matter (e.g., "Are calories from fat different than calories from sugar?" "Are all fats bad for you?"). Introducing questions before students have mastered content allows for assessment of prior knowledge, identification of misconceptions, and provides a baseline that can be revisited after the unit to assess growth in students' analytical sophistication on the issue.

6. **Provide Formal Instruction.** Present subject matter in a variety of ways (e.g., direct teaching, laboratory investigations, internet/textbook research, class discussions, guest speakers, field trips, films) in order to ensure content coverage and student engagement.

7. **Incorporate Group Activities**. Allow students to investigate issues in small groups, emphasizing the analysis of evidence, reinforcement of content matter, monitoring of understanding, consideration of divergent viewpoints, and full participation of all students (see section on "Cooperative Learning").

8. **Provide Guidance in Evaluating Primary and Secondary Sources.** Discuss the importance of identifying bias and provide tools for assessing trustworthiness of research sources. (see section on "Evaluating Sources").

9. **Assess Knowledge and Reasoning.** A variety of "products" that allow students to demonstrate learning can be developed. These include group presentations, class debates or town hall forums, posters, papers, videos, written tests, and so on … (see section on "Assessment"). Assessment should include content understanding as well as its reasoned application to the issue at hand.

10. **Have Fun!** SSI provides opportunities for creativity, engagement, and exploration by students and teachers. Once you have established a "rhythm" for classroom discussion and a level of comfort with your role as facilitator of learning, you will reap the benefit of seeing your students embrace science learning and applying it to contexts that prepare them for informed and engaged citizenship.

Key Classroom Strategies for SSI Implementation

In order to facilitate the development of your own SSI units, we have included the following sections to highlight some key classroom strategies and skills that can be developed and nurtured to promote success in an SSI curriculum.

Argumentation

For many, the term *argument* has a negative connotation, as it is often associated with quarrels or disputes. Yet in science, argumentation is quintessential to the process of intellectual engagement, formulating ideas and conveying them to others using evidence (Osborne, Erduran, and Simon 2004). Argumentation is a pattern of reasoning that allows students to organize findings, advance claims, and identify alternatives. Through sharing of claims and evidence, students process information, learn new ideas, and ultimately test the validity of their findings and conclusions. This type of discourse models the work of real scientists and helps in the negotiation of real-world science-related issues (Driver, Newton and Osborne 2000). Argumentation is foundational not only in science, but also in language arts and mathematics. According to the *NGSS*, "students are expected to engage in argumentation from evidence; construct explanations; obtain, synthesize, evaluate, and communicate information; and build a knowledge base through content rich texts across the three subject areas" (NGSS Lead States 2013, Appendix D, p. 22). The key to teaching argumentation is helping students to argue using evidence, and not simply opinion or belief. This does *not* imply that students need to avoid arguments that are rooted in emotive convictions, such as caring, empathy, and the like; only that such passions are accompanied by the use of scientific evidence. This can be quite challenging, especially when the issue being argued is personally meaningful to students as they instinctually rely on emotional responses without consideration of evidence. Instruction and practice are required to make evidence-based argumentation the norm in one's classroom. SSI is a powerful tool for encouraging argumentation skills as controversial issues provide an ideal context for students to wrestle with, negotiate, and debate their ideas.

Here are some tips, drawn from our teaching experience and research on effective classroom implementation of SSI (Zeidler, Applebaum, and Sadler 2011), for scaffolding argumentation skills while implementing SSI:

1. Create an intellectually safe classroom environment.

   Students will be more apt to share their ideas and take intellectual risks if they feel they are respected and their ideas are important. Here are some strategies to accomplish this:

- Model active listening skills in the classroom by being attentive to all ideas and rephrasing them to clarify and emphasize their importance.
- Discuss the difference between challenging someone's ideas and attacking them personally before allowing students to engage in any type of debate or discussion forum.
- Avoid mockery of any kind—what seems like a funny, flip comment might be interpreted as hurtful by the student.
- Encourage students to discuss ideas in small groups before sharing with the whole class. This can boost confidence and make students feel more secure.
- Remember that the same rules that apply in the classroom also apply online—if you use discussion boards, blogs, and so on, make sure that students are questioning ideas and not attacking others.

2. Provide students with frameworks for argumentation.

   Helping students to structure their arguments and keep track of others' arguments can be invaluable in promoting evidence-based reasoning. Some suggestions follow:

   - Use T-Charts, either individually or in small groups, to help students organize their arguments ("agree" and "disagree") on an issue. You can also divide the room into agree/disagree sides (or along a spectrum line) and have students discuss their ideas before or after recording them. While the students may have little prior knowledge at the start of an activity, these methods help them (and you) to identify gaps in their knowledge, providing motivation for further research.
   - Emphasize (and re-emphasize!) claims and evidence —for each claim a student makes, remind them that they must provide their evidence for that claim, and be able to identify the source of the evidence (i.e., from observation, experimentation, online/book research). This can be recorded on a chart with the column headings "Claims," "Evidence," and "Sources."

Claims	Evidence	Sources
"Wind power is the fastest growing source of electricity"	"There has been a ___% percent increase in wind farms in the U.S. over the last 5 years"	"According to __________ published in/at _________"

- Encourage students to anticipate counterarguments by asking, "What would someone who disagrees with you say about that...and why?" You can then ask, "And what would you say in response?" to help them structure a rebuttal.
- Provide students with sentence frames such as these:
    - I agree/disagree with ... because ...
    - I observed/researched ... and found ...
    - Our evidence was similar to ... but we concluded that ... because ...

3. Provide multiple means of practicing, expressing, and assessing argumentation skills.
    - Students can engage in scientific argumentation during various types of activities including hands-on investigations, library/internet research, and/or interactive debates. The list of possibilities is endless, but the key is to emphasize the use of evidence to back up their arguments.
    - Remember that argumentation does not have to be oral; writing, pictures, and drawing also serve as important expressions of reasoning. Some creative modes of expressing argumentation include role-play, cartooning, online discussion boards, letter writing, e-stories, and many others. See the section on "Assessment" for more ideas.

You will find that thoughtfully incorporating argumentation strategies in your science classroom will lead to more active engagement of students with subject matter, and develop a sense of personal identity to science, as well as a sense of community amongst themselves. These strategics also help students to challenge their own thinking, as they are often compelled to evaluate their own thinking skills thoughout the process. These outcomes are consistent with increasing the public understanding of science through sociocultural engagement.

References

Driver, R., P. Newton, and J. Osborne. 2000. Establishing the norms of scientific argumentation in classrooms. *Science Education* 84: 287–312.

NGSS Lead States. 2013. *Next generation science standards: For states, by states.* Washington, DC: National Academies Press.

Osborne, J., S. Erduran, and S. Simon. 2004. Enhancing the quality of argumentation in school science. *Journal of Research in Science Teaching* 41: 994–1020.

Zeidler, D. L., S. M. Applebaum, and T. D. Sadler. 2011. Enacting a socioscientific issues classroom: Transformative transformations. In *Socio-scientific issues in the classroom*, ed. T. D. Sadler, 277–305. Netherlands: Springer.

Assessment

Assessment is a critical practice for informing teaching and learning. Ensuring that formative and summative assessments align with objectives and are consistent across students is essential. SSI provides particularly exciting opportunities for multiple forms of assessment, as we can examine and better understand students' conceptual understanding of science content and process skills, as well as evidence-based reasoning.

Preliminary questions to be considered when designing assessment for SSI are:

- What is our objective?
- How will we know when it is achieved?
- What are our judgment criteria?
- What alternatives exist to traditional pen-and-paper examinations that may tap SSI -related skills?

In assessing the success of SSI implementation, it seems apparent that identifying means of evaluating evidence-based reasoning is important, since the ability of students to synthesize and integrate data from their own investigations or from other reliable sources and utilize it as evidence to support their claims is a major goal of SSI. In assessing students, criteria may include using evidence, rather than opinion, to support their claim, integrating science content into their arguments, and recognizing that some sources of data (such as books, reliable websites, their own observations) are more reliable than others (such as commercials, friends, or websites with a particular bias). More sophisticated levels of understanding might include the ability to evaluate the differences in quality among scientific studies. Some dimensions of evidence-based reasoning that we may wish to consider are:

1. Use of Evidence: Is the student using empirical data (as opposed to opinion) to make his or her argument?
2. Source of Evidence: Did the student evaluate the reliability of the source of evidence? (i.e., a research journal vs. a TV commercial)

3. Quality of Data: Did the student evaluate the quality of the study cited and/or confirm findings with other studies? (i.e., what methods were used? Have these findings been confirmed by others?)
4. Science Content: Is the student applying science content understanding to their interpretation of the data?

A sample rubric that could be utilized to assess evidence-based reasoning is shown in Table P2.1.

TABLE P2.1.

Rubric for Scoring Evidence-Based Reasoning

	Early (1 pt)	Emerging (2 pts)	Sophisticated (3 pts)
Use of Evidence	Student uses opinion without evidence or inaccurate evidence to back their claims	Student uses tenuous or incomplete evidence to back claim	Student demonstrates complete and accurate use of evidence to back claims
Source and Quality of Evidence	Student does not recognize or distinguish between reliable and unreliable sources of data	Student demonstrates some effort in evaluating the quality of the sources of data	Student thoughtfully evaluates the quality and reliability of the data and can articulate process
Science Content Understanding	Student demonstrates minimal understanding of science content	Student demonstrates a moderate degree of understanding of science content	Student demonstrates strong understanding of science content and consistently applies it to their argument

There are many "products" that can be used for evaluation while implementing SSI. Students should be given many opportunities to demonstrate their understanding of science content and their ability to formulate claims using evidence. Some products for evaluation include

- poster board presentations,
- position papers,
- brochures,
- letters to the editor (business, congress, senator, and so on),

- online discussion boards,
- storyboards,
- debates or town hall meetings,
- Performance of investigative or inquiry research or survey,
- PowerPoint presentations,
- videos,
- dramatic reenactments,
- public service announcements (PSA),
- cartoons or comic strips, and
- participation in group discussions.

FIGURE P2.1.

Examples of Poster Boards Created by High School Biology Students
(photos courtesy of Scott Applebaum)

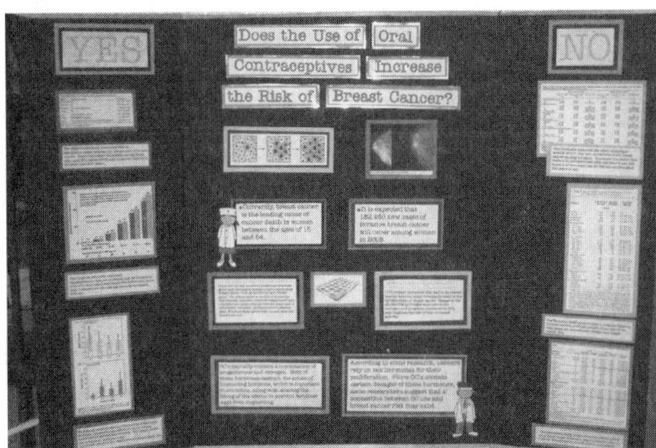

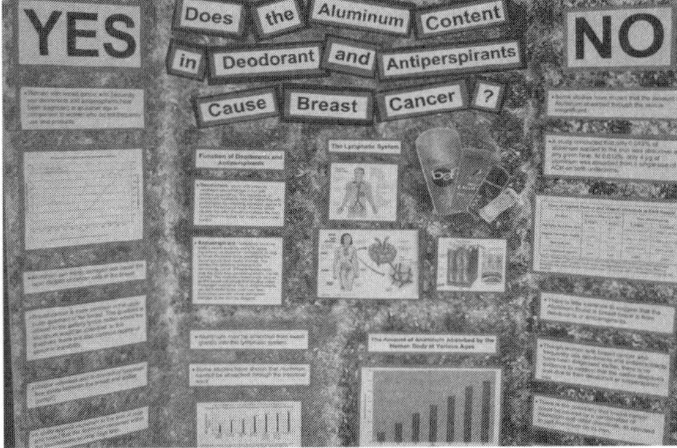

Because debate and town meetings are so frequently utilized in SSI, we have included a sample debate rubric below in Table P2.2. This rubric can be customized to align with the particular topic and/or stakeholders involved in the debate.

TABLE P2.2.

Sample Debate Rubric

	Beginning (1 pt.)	Developing (2 pt.)	Average (3 pt.)	Emerging (4 pt.)	Accomplished (5 pt.)
Opening Statement Strength of claims and evidence	The presenters fail to state claims in their statement.	The presenters include claims but use only opinion to support them.	The presenters include claims but use a combination of opinions and empirical evidence to support them.	The presenters use empirical evidence to support their claims but the evidence is tenuous or erroneous.	The presenters use accurate and valid evidence to support their claims.
Clarity	The presenters are inaudible or unclear and make no attempt to engage the audience.	The presenters are audible but the arguments are confusing and/or convoluted.	The presenters communicate their arguments but are not persuasive.	The presenters communicate some of their arguments in a clear, persuasive manner.	The presenters communicate all of their arguments in a clear, persuasive manner.
Cross-Examination Strength of questions	No questions are asked.	Questions do not relate directly to opening statements.	Questions relate to opening statements but simply ask for reiteration of facts rather than probe for contradiction.	Questions probe for contradictions but fail to follow-up when new information is learned.	Questions probe for contradictions and evolve as new information is learned.
Clarity	The questions are inaudible or unclear.	The questions are audible but are confusing and/or convoluted.	The questions are audible but the team has not organized the order of questions or questioners.	The questions are audible and clear, but the order of questions is illogical and confusing.	The questions are audible, clear, and are asked in an orderly, logical order.

	Beginning (1 pt.)	Developing (2 pt.)	Average (3 pt.)	Emerging (4 pt.)	Accomplished (5 pt.)
Rebuttal Strength of arguments and evidence	The presenters fail to present a rebuttal.	The presenters attempt to rebut but they are unable to cite evidence.	The presenters attempt to rebut but use a combination of opinions and empirical evidence to do so.	The presenters rebut using empirical evidence to support their claims but the evidence is tenuous or erroneous.	The presenters use accurate and valid evidence to form their rebuttal.
Clarity	The presenters are inaudible or unclear and make no attempt to engage the audience.	The presenters are audible but the rebuttal is confusing and/or convoluted.	The presenters communicate their rebuttal but are not effective in defending their stance.	The presenters communicate some of their rebuttal in a clear, effective manner.	The presenters communicate their rebuttal in a clear, effective manner.
Closing Statement Strength of arguments and evidence	The presenters fail to summarize claims in their statement.	The presenters summarize claims but use only opinion to support them.	The presenters summarize claims but use a combination of opinions and empirical evidence to support them	The presenters use empirical evidence to support their claims but the evidence is tenuous or erroneous.	The presenters summarize their claims using accurate and valid evidence to support them.
Clarity	The presenters are inaudible or unclear and make no attempt to persuade the audience.	The presenters are audible but the arguments are confusing and/or convoluted.	The presenters communicate their arguments but are not persuasive.	The presenters communicate some of their arguments in a clear, persuasive manner.	The presenters communicate all of their arguments in a clear, persuasive manner.
Total Score					

Scoring: Scale of 1 (low) to 5 (high)
Maximum Score: 40 points per team

Several other assessment tools are included in this book within each of the sample units. It is essential to remember that there are a variety of tools at your disposal to make assessment and evaluation effective and enjoyable. It is also important that students are aware of the assessment schemes that will be used. Finally, it may prove to be an interesting exercise to involve students, at times, in the refinement of possible class rubrics that could be used for SSI investigations, thereby fostering authentic learning experiences.

Evaluating Sources of Information

Today's students are products of the "Information Age," with unprecedented, round-the-clock access to information from books, newspapers, magazines, journals, television, radio, and of course, the internet. Most students are able to access information from the palm of their hands, yet few have been instructed in strategies to evaluate the reliability of the sources. The ability to evaluate sources of information is critical in understanding the complex and often conflicting viewpoints advanced in the negotiation of societal issues. While it may be tempting to only direct students to resources that are considered "objective," we believe it is important to allow students to experience information in way that emulates their exposure in the real world. To that end, students should *not* be instructed to avoid highly opinionated or biased sources, but rather, taught to identify the bias and use the material to understand the particular perspective being advanced. If students can recognize material designed to persuade, advocate, or advance a particular agenda, they will be well-positioned to weigh it against information from more varied and possibly reasoned sources, acting on it from a more balanced perspective.

There are many guides for evaluating sources of information. One useful framework is referred to as C.A.R.S. (Harris 2000), which is an acronym for "Credibility," "Accuracy," "Reasonableness," and "Support." The framework can be summarized as follows:

Credibility

- What are the author's credentials (e.g., position, education, reputation)?
- Is there evidence of quality control such as peer review?

Accuracy

- Is the material current and comprehensive?
- Is the material written for the intended audience and purpose?

Reasonableness

- Does the material seem fair and objective, or highly biased?
- Is the tone consistent and moderate?

Support

- Is the source of information documented through references/bibliography?
- Can the information be corroborated?

It would behoove teachers to provide students with a collection of resources on the same topic, representing a range of reputability, and encourage students to evaluate the materials' trustworthiness according to the C.A.R.S. framework. Class discussion should focus on recognizing the biases and identifying the intended purpose (if any) of the various materials. The more practice students have in this skill, the more prepared they will be for evaluating information that influences their decision making in real-world contexts.

One final aspect of assessing sources deals with the nature of scientific information. Students are often surprised to see that reputable sources of scientific information may report conflicting findings from similar studies. This may lead students to think that scientific research, no matter how well designed, is untrustworthy. Teachers need to stress that scientific information, while durable, is tentative and always subject to scrutiny, review, challenges, and change, in light of more compelling evidence due to the emergence of new findings and technologies. If students can understand that conflicting scientific evidence is, at least in part, due to the nature of scientific inquiry, they may come to appreciate that their research efforts have made them privy to the ongoing dialogue among scientists that advances our understanding of the world.

Reference

Harris, R. 2000. *WebQuester: A guidebook to the web.* Boston: McGraw-Hill.

Cooperative Learning

Science, like any discipline, requires collaboration for its advancement. Helping students to discover and debate together enhances the learning process and prepares them for participation in society. Social skills are key factors in career advancement as well as interpersonal relationships, yet many social skills are surprisingly lacking, even in older students. Cooperative learning (Johnson and

Johnson 2002; Slavin 1980) is a research-based approach to classroom organization that fosters cooperative skills. It is a particularly effective way to promote understanding among diverse populations as students with different backgrounds and skills learn from each other and grow together (Putnam 1997). SSI lends itself to the implementation of several cooperative learning strategies as it is a highly interactive framework that is enhanced by students' ability to communicate freely and productively together.

Cooperative learning is comprised of five main elements: (1) individual and group accountability; (2) positive iinterdependence; (3) face-to-face interaction; (4) promotion of social skills; and (5) group processing. Each of these elements can be integrated into the negotiation of SSI by using the following tips:

1. Assign and rotate team roles, such as recorder, reporter, timekeeper, materials manager, principal investigator, during activities to ensure full participation.
2. Ensure that the group task (i.e., book or online research, laboratory investigation, presentation) requires the input of all students; in other words, make each role essential.
3. Assess students on both individual work and group work (see Section on Assessment, p. 37).
4. Consider distributing "C-Points" to teams for demonstrating desired social skills. C-Points can consist of a tally of points in order to earn some goal, such as a privilege, or simply points the team earns toward its group project grade. Here is a list of some "C" skills that can be targeted:
    a. Cooperation
    b. Collaboration
    c. Communication
    d. Consensus-Building
    e. Care of Materials
    f. Completion

    Be sure to model and describe the social skill you are targeting so that students understand expectations.

5. Process group activities by having students provide input on their group's performance including areas of strength and challenges. Highlight the processes groups followed to work through challenges.
6. Emphasize team building and class building (Kagan, Kagan, and Kagan 1997) in order to foster trust and positive communication among team members.

Some common cooperative learning techniques used to organize and manage lessons are:

- *Think-Pair-Share:* Allow students to individually reflect on a question posed, then give them time to discuss with a classmate before engaging in whole-class discussion.
- *Jigsaw:* Have team members become experts on a certain research topic by working with members from other teams and then return to teach what they've learned to their original team; students learn by teaching and the team is reliant on each student becoming an expert.
- *Numbered Heads Together:* Assign a number (or have students number off) in their groups. Pose a problem or question to students, have groups work on a solution together, and then call a random number to report to the class. This strategy encourages the group to ensure that all members understand the concepts.

As with any instructional strategy, cooperative learning should not be used to the exclusion of other forms of knowledge construction. After all, students will need to learn how to take responsibility for their own learning too. However, the ability to work collectivey through problems is a lifetime skill that will aid students in the practice of science not only with SSI, but also in the practice social justice and the development of character.

References

Johnson, D. W., and R. T. Johnson. 2002. Cooperative learning and social interdependence theory. In *Theory and research on small groups*, ed. R. S. Tindale, 9–35. New York: Plenum Press.

Kagan, L., M. Kagan, and S. Kagan. 1997. *Cooperative learning structures for teambuilding.* San Clemente, CA: Kagan Cooperative Learning.

Putnam, J. W. 1997. *Cooperative learning in diverse classrooms.* New Jersey: Prentice Hall.

Slavin, R. E. 1980. Cooperative learning. *Review of educational research* 50 (2): 315–342.

Guide to Sample Units

In Part 1, we attempted to "make the case" for SSI in the K–12 classroom by presenting the framework, rationale, and extensive research base that supports it. If we've been successful, you are probably anxious to get started implementing SSI in your classroom. Although there is no set "formula" for constructing SSI curricula, we wished to provide the reader with recommendations for developing SSI curriculum as well as a series of sample unit plans that reflect the key SSI features including: (1) Consideration of a controversial, socioscientific issue; (2) Deliberation of the moral/ethical dilemmas involved; (3) Opportunities for argumentation and discourse; and (4) Context for rigorous science content and process skills.

The units included in this book were based on projects developed by several thoughtful, creative teachers participating in a course on socioscientific issues at the University of South Florida. We are grateful for their efforts and willingness to contribute their works to this book. The units have been revised to reflect a consistent format that includes the following:

- Unit Overview: a brief synopsis of the unit
- Key Science Concepts: key terms and ideas to provide the teacher with quick reference
- Ethical Issues: brief descriptions of the ethical issues/dilemmas involved in each unit
- Science Skills: science process skills reflected in the unit
- Grade Levels: a grade range–based subject matter, reading level, and standards
- Time Needed: estimate of time for unit based on 50-minute periods
- Lesson Overview: list of lessons with brief descriptions
- Background on the Issue: summary of the history and controversy surrounding the issue
- Connections to *NGSS*: alignments with the *Next Generation Science Standards*
- Accommodations for Students With Disabilities: ideas for adapting units for students with visual, hearing, motor-orthopedic, learning, and emotional disabilities
- Resources for Teachers: books and websites for additional reading/reference

Each unit consists of several flexible lesson plans, many of which can be utilized as templates for use with other issues. The lesson plans consist of the following sections:

- To the Teacher
- Objectives
- Time Needed
- Materials
- Accommodations for Students With Disabilities*
- Procedure
- Closure
- Assessment

*Included in individual lesson plans for elementary level units due to extensive hands-on activities.

The seven units included in this section reflect a range of disciplines, grade levels, and instructional strategies. They are meant to demonstrate the types of challenging, thought-provoking questions that can be addressed through SSI and the varied means of engaging students in contextualized science learning. They are by no means an exhaustive collection of SSI lessons but rather, samples to help inspire and guide the novice SSI practitioner. After trying some of these units in the classroom and reviewing the introductory sections of Part 2, we hope that you will feel prepared to embark on the adventure of designing your own units!

List of Sample Units

Elementary School Level

Unit 1: Food Fight

Should schools charge a "fat tax" for unhealthy foods?

Unit 2: Animals at Work

Should animals perform in circuses?

Middle School Level

Unit 3: A Need for Speed?

Should speed limits be lowered to reduce traffic fatalities?

Unit 4: Space Case

Do humans have the right to colonize and use resources on extraterrestrial planets?

High School Level

Unit 5: A Fair Shot?

Should Gardasil vaccines be mandatory for all 11–17-year-olds?

Unit 6: "Mined" Over Matter

Should rare Earth elements be mined in the United States?

Unit 7: "Pharma's" Market

Should prescription drugs be advertised directly to consumers?

Unit 1

Food Fight

Should schools charge a "fat tax" for unhealthy foods?

Unit Overview

During this unit, students will learn the essentials of good nutrition, including the identification and functions of major nutrients, how healthy foods benefit the body, and strategies for making healthy food choices. By engaging in the controversial question of whether school cafeterias should charge more money for unhealthy foods, students will use research and experimentation on nutrition to derive evidence to support their claims about "What is a healthy food?" "Who decides what's healthy for me?" "Should advertisers target children in food ads?" and "How do we best deal with the problem of childhood obesity?"

Key Science Concepts

Nutrition, Digestive System Form and Function, Actions of Preservatives, Healthier Food Alternatives

Ethical Issues

Personal Choice, Government Control, Paternalism, Economic Disincentives, Advertising Targeted Toward Children

Science Skills

Observing, Comparing and Contrasting, Recording and Analyzing Data, Formulating Arguments From Evidence, Communicating Results, Inferring

Grade Level

Elementary

Time Needed

This unit includes eight lessons that can be adjusted to the pace and schedule of your class.

Lesson Sequence

- **Lesson 1.** Initial Debate: Should Schools Charge More Money for "Unhealthy" Foods?
- **Lesson 2.** Introduction to the Food Groups: My Healthy Plate
- **Lesson 3.** How Fresh Is Your Food? The Effects of Preservatives
- **Lesson 4.** The Six Nutrients Mystery Matching Game and Reading Labels Carousel
- **Lesson 5.** Digestion: The Path to Good Nutrition
- **Lesson 6.** Find the Fat … and Cheeseburgers on Trial!
- **Lesson 7.** Kids as Consumers: Evaluating and Creating Food Commercials
- **Lesson 8.** Final Debate and Letter Writing: Should Schools Charge More Money for "Unhealthy" Foods?
- **Optional Lesson.** Field Trip to Supermarket and "Eat This, Not That" Slide Show

Background on the Issue

Childhood obesity has increased significantly in the United States over the last several decades, with recent statistics suggesting that more than 1 in 6 American children are obese (CDC 2012). Obesity in children and adults is a serious matter as it leads to many related health problems, including diabetes and heart disease. The causes of increased obesity in children include the fact that many children lead rather sedentary lifestyles with limited physical exercise or time spent playing outdoors. Furthermore, the "fast food nation" phenomenon continues to exist, with many American families eating out at fast food restaurants or eating "quick fix" meals several times a week.

1

Under the new Healthy, Hunger-Free Kids Act signed by President Obama in 2011, school lunches will need to adhere to strict guidelines regarding the proportions of fruits and vegetables made available at lunch, as well as the requirement of low fat and low sodium (Wood 2012). While schools will receive additional funds to implement the changes, it is not clear that students will actually choose the healthy foods. Denmark became the first country to charge a national "fat tax" for unhealthy foods (Zafar 2011), and many others, including the United States, are considering following suit. Having a "disincentive" to purchase unhealthy foods is a controversial issue, touching on issues of individual choice, government control, and the question of "what is healthy?" This is a timely, relevant issue that is sure to pique students' interests and empower students to use scientific knowledge to be informed consumers in regard to their own health, as well as knowledgeable citizens who can contribute thoughtfully to social policy issues.

Connecting to *NGSS*

K-LS1-1. Use observations to describe patterns of what plants and animals (including humans) need to survive.

4-LS1-1. Construct an argument that plants and animals have internal and external structures that function to support survival, growth, behavior, and reproduction.

5-PS3-1. Use models to describe that energy in animals' food (used for body repair, growth, motion, and to maintain body warmth) was once energy from the sun.

Disciplinary Core Ideas

LS1.C: Organization for Matter and Energy Flow in Organisms. All animals need food in order to live and grow. They obtain their food from plants or from other animals. Plants need water and light to live and grow. (K-LS1-1)

LS1.A: Structure and Function. Plants and animals have both internal and external structures that serve various functions in growth, survival, behavior, and reproduction. (4-LS1-1)

Accommodations for Students With Disabilities

Visual Impairments: Enlarge all handouts and/or make readings available on a large-screen monitor for students with low vision; Make all materials tactile using materials such as school glue (becomes "raised" when dried), Wikki Stix (wax-covered string), puffy paint, or by use of a Braille labeler; allow students to handle materials in advance of class activities.

Hearing Impairments: Give all instructions in both oral and written (or pictorial) formats; use strategies such as "pass the microphone" (only the student holding the "microphone" can speak) so that hearing impaired students can easily identify the speaker; use visual cues (such as lights on/lights off) for lesson transitions.

Learning Disabilities: Use graphic organizers such as T-Charts, Venn Diagrams, and Concept Maps to help students organize and retain information; Pre-teach vocabulary and offer students multiple opportunities to demonstrate understanding; use "wait time" for all students.

Motor/Orthopedic Impairments: Make sure that all work areas are at an appropriate height for students who utilize wheelchairs; use large materials for students with fine motor challenges; Velcro can be added to items to help them stay on desks, and can also be put on gloves to help students grasp items; do not assume what a student can't do … speak to the student about their ideas for how to accomplish a task.

Emotional Disabilities: Anticipate challenges in dealing with cooperative skills such as consensus building, respectful listening and discourse, and collaboration; Assist students and teammates with strategies to ensure all ideas are heard, such as "mirror listening," and "one idea add-ons"; reiterate that there are no right answers to SSI questions … students can (and will) disagree … they just need to justify their opinions with evidence.

Resources for Teachers

Nutrients

- *www.choosemyplate.gov*
- *www.editorsweb.org/nutrition/essential-nutrients.htm*
- *www.ncagr.gov/cyber/kidswrld/index.htm*
- *www.nourishinteractive.com*
- *http://kids.usa.gov/grown-ups/exercise-fitness-nutrition/index.shtml*

Digestion

- *www.neok12.com/Digestive-System.htm*
- *http://kidshealth.org/kid/htbw/digestive_system.html*

References

Centers for Disease Control (CDC). 2012. Prevalence of obesity in the United States, 2009–2010. *www.cdc.gov/obesity/childhood/index.html*

Wood, S. 2012. Students to see healthier lunches under new USDA rules. *http://usnews.msnbc.msn.com/_news/2012/01/25/10234671-students-to-see-healthier-school-lunches-under-new-usda-rules*

Zafar, A. 2011. Denmark institutes first ever "fat tax." *Time Newsfeed.* September 30. *http://newsfeed.time.com/2011/09/30/denmark-institutes-first-ever-fat-tax*

Note: This unit is based on a project contributed by Christina Cullen and Hayley Sweet.

Lesson 1

Initial Debate

Should Schools Charge More Money for "Unhealthy" Foods?

To the Teacher

In an effort to encourage healthier eating, some countries, including the United States, are considering charging more money for unhealthy foods. This is often referred to as a "fat tax." In this activity, your students will discuss the kinds of foods they like to eat and then informally debate whether they think school lunches that include "unhealthy" foods should cost more money than "healthy" lunches.

Objective

Students will be able to informally assess their initial opinion of an SSI Question, hear other students' opinions, and begin to clarify the meaning of "healthy" vs. "unhealthy."

Time Needed

One class period

Materials

Materials depend on preferred options for voting: *For Paper Voting:* 1 sticky note and pencil per student; *For "Mimio" Electronic Voting:* Mimio and Mimio Vote; *For "Decision Line" Voting:* 1 long strip of masking tape on the floor with one side designated as "Yes" and the other side designated as "No"; *For "heads-down" vote:* no materials needed.

Procedure

1. The teacher will pose the SSI Question, "Should school cafeterias be allowed to charge more money for unhealthy foods?"
2. Using one of the voting procedures described above in the "Material Section," students will vote either "yes" or "no."
3. Next, the teacher will graph the results. If using sticky notes, simply post the notes on the board to form a pictogram. For others, create a bar graph.
4. Ask several students from each category to explain why they feel the way they do.
5. Using Mix-Pair-Share, have students respond to the following questions:
   - What makes certain foods healthy or unhealthy?
   - What would your "perfect" cafeteria lunch look like?
   - Why do you think the issue of charging more for "unhealthy" lunches has come up?

Closure

Explain to students that over the next several lessons, they will be doing different activities and experiments to learn about foods so that they can have more information about the "fat tax" question.

Assessment

Have students complete the "Fantasy Lunch" T-Chart which will be used as a pre-assessment to the students' thoughts on the SSI question.

Accommodations for Students With Disabilities

Visual Impairments: Make sure that all graphs displayed are large. Use of the sticky note method for voting makes it tactile as well; if "line voting" is used, be sure aisles are clear; create tactile T-chart by going over the line markings with glue and allowing it to dry (Wikki Stix can be used for this, too). If Mimio Vote is used, be sure that student can distinguish between the "yes" vote (a check mark) and a "no" vote (an "x" mark). You can make the buttons more tactile by using "Wikki Stix" (wax-covered string) or with a small amount of clay.

Hearing Impairments: Put the question, "Should schools charge more money for unhealthy foods?" in writing on the board; stop and start group discussions by use of lights on/lights off signals; review vocabulary with student in advance and put main vocabulary words in writing because many words, such as "fat" and "tax," can sound similar.

Learning Disabilities: Multiple formats for creating bar graphs can be used to reinforce the meaning of the graphs. For example, having students line up according to their votes, posting written sticky notes, and creating a digital display can clarify the use and interpretation of bar graphs.

Motor/Orthopedic Impairments: Do not assume what a student can or can't do. In the case of voting, a student who does not have use of his or her hands may be able to use feet or use a mouth stick to press buttons. Ask the student in advance for input; if "line voting" will be used, be sure that the student has clear access or if preferred, make the student's desk the position of his or her preferred vote and have other students come to him or her.

Emotional Disabilities: Be clear with all students that the question of whether schools should charge extra money for unhealthy foods is an opinion question, and that there is no one right answer, as long as students can express an opinion (and they will soon be able to back it up with facts); accept all answers; remind students that they will have other opportunities to express their opinions during this unit.

Name: ______________________________ Date: ____________

"Fantasy Lunch" T-Chart

Fantasy Lunch Items	Why Included

1. **I think my "Fantasy Lunch" is:**

   (Circle one)

   Very healthy.

   A little healthy.

   Healthy and unhealthy.

   A little unhealthy.

   Very unhealthy.

2. **I think school cafeterias should charge:** *(Circle one)*

   The same price for all lunches.

   More money for healthy lunches.

   More money for unhealthy lunches.

   No money for lunch. I have another idea: ____________________

3. **My thoughts …**

Lesson 2

Introduction to the Food Groups

My Healthy Plate

Topic: Nutrition
Go to: *www.scilinks.org*
Code: SSI001

To the Teacher

This activity is focused on building student knowledge on the major food groups. In 2011, the USDA replaced the familiar food pyramid concept with a "My Plate" approach to food group selection. In this way, children learn to think of their plates as an opportunity to build a healthy meal that includes the five food groups (fruits, vegetables, proteins, grains, and dairy.) The "My Plate" approach also visually communicates recommended proportions of food. (For example, half of the plate should be filled with fruits and vegetables). During this activity, children will hypothesize about the ways to sort foods, review the "My Plate" food groups and proportions, re-sort their food, and then create their own "healthy plate" representing a healthy meal.

Objectives

Students will be able to name the major food groups: protein, dairy, fruits, vegetables, grains, and extras, and will be able to correctly sort foods into the appropriate category. Additionally, they will become familiar with the dietary recommendations of the USDA and be able to plan a meal that approximates the suggested proportions of the food groups.

Time Needed

One class period

Materials

- Food magazines or supermarket circulars with pictures of various foods
- Scissors

- Glue
- Paper plate (2 per child; includes 1 for homework)
- My Plate Handout

Procedure

1. Divide students into groups of three to four students. In groups, have students cut out pictures of various foods from the food magazines or supermarket circulars. After cutting, have each group sort and place their pictures into categories of their choosing (do not provide students with a food guide). Once students have finished classifying foods, have each group present how they classified their foods to the rest of the class. After all groups have presented, ask students for ways to make classification easier and more consistent. After a brief discussion, begin to introduce the concept of the food guide. Distribute the "My Plate" Food Guide (p. 64) to each group and have each group reclassify their foods according to this resource.
2. Students are now challenged to create a "meal" on their paper plate by gluing the pictures onto their plate and using the "My Plate" handout as a guide.
3. Have students evaluate each other's meals by checking to make sure that all five groups are represented and are in correct proportions.

FIGURE 1.1.

A Sample "My Healthy Plate"
(photo courtesy of Sami Kahn)

Closure

1. Ask, "Can you name the five food groups?"
2. "How does the "healthy plate" help us to know which foods to choose?
3. Are "extras" always bad? Explain. (OK in moderation if diet is primarily food groups with high amount of fruits and vegetables)

Assessment

1. Students should correctly complete their cutting and gluing activity (Note: As an extension, they may add foods into each category.)
2. Students will bring home a "blank" plate and draw in foods that they eat by putting them in the proper groups.

Scoring Rubric

1 pt.	2 pts.	3 pts.	4 pts.	5 pts.
Plate shows examples of only one food group	Plate shows examples of two food groups	Plate shows examples of three or four food groups	Plate contains examples of all five food groups but proportions are highly inaccurate	Plate contains examples of all five food groups in approximate proportions

Extensions

Students can reinforce their understanding of the food groups by engaging in the USDA's online game, "Blast Off," found at *www.fns.usda.gov/multimedia/Games/Blastoff/BlastOff_Game.html*. In this game, students select foods that will give them the proper balance of nutrients and calories sufficient to "blast off" into space.

Accommodations for Students With Disabilities

Visual Impairments: Be sure to make large-print handouts available; use plastic play foods for blind students to sort; food photos can be made tactile using school glue and sand or salt; use Braille labels if needed.

Hearing Impairments: Introduce students to the important vocabulary such as "nutrition," "grains," "proteins," and so on in advance so that students are familiar with them; give instructions in written and verbal formats; remind teammates and classmates to face student when speaking to him or her.

Learning Disabilities: Give students opportunities to sort other items (such as coins or beans) in advance so that the skill is familiar; have students create, "healthy food rings books" by gluing the "My Plate" categories on index cards, punching a hole in, and hanging them on a ring as a reminder for student.

Motor/Orthopedic Impairments: Use adaptive scissors for cutting or allow student to select precut photos; use large squeeze bottles for glue; remind all students that it's OK if things get messy, as long as they clean up (to help student with motor skills difficulties to feel less pressured).

Emotional Disabilities: Remind teammates that all ideas regarding food sorting are to be considered; there are many ways to sort the foods correctly; encourage all teammates to participate in the food sorting presentation; allow student to be a "plate checker" if he or she is having difficulty with the activity—this may give the student a feeling of security and confidence to help him or her finish the job.

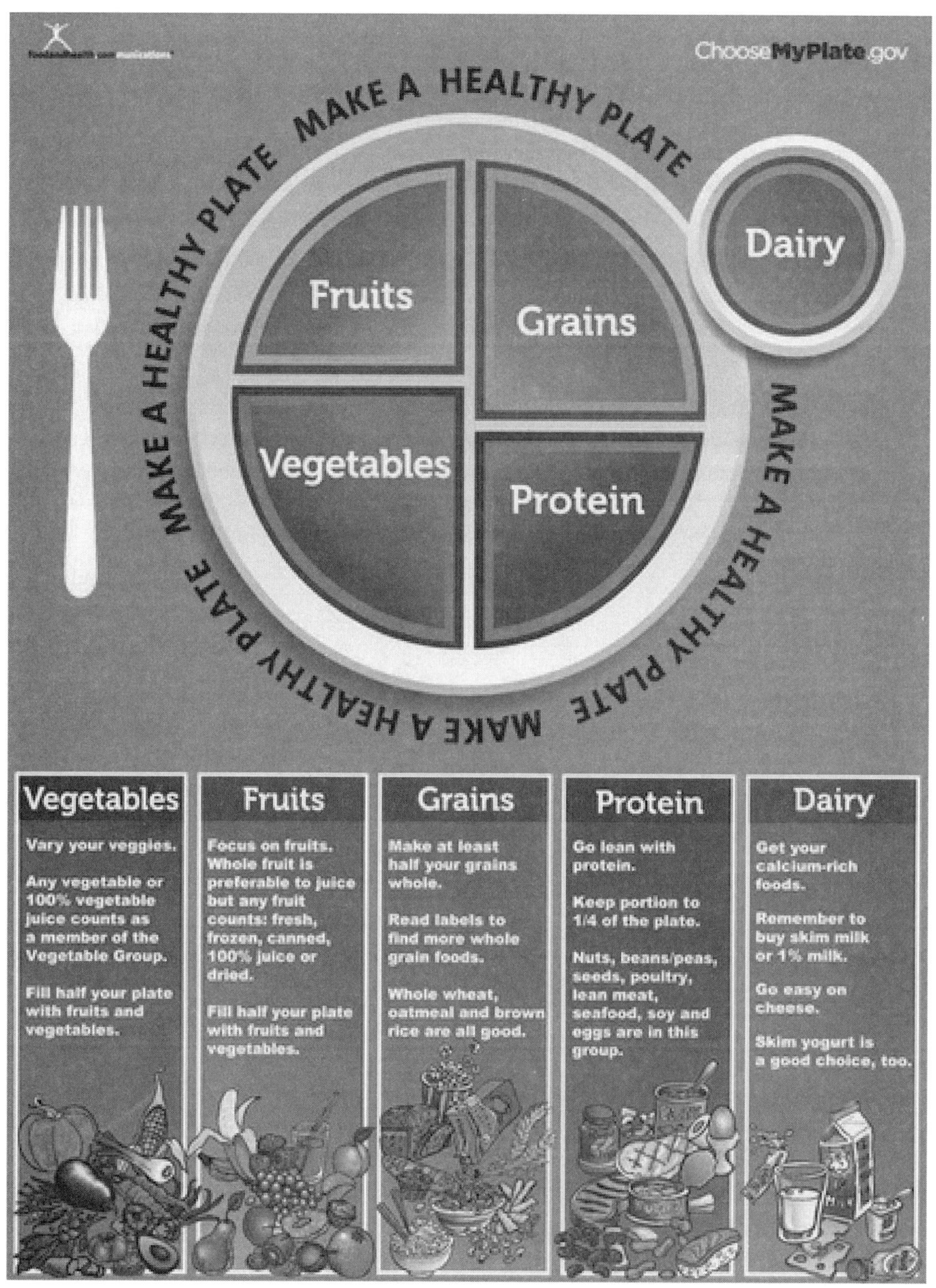
ChooseMyPlate.gov
MAKE A HEALTHY PLATE
Fruits
Grains
Vegetables
Protein
Dairy
Vegetables
Vary your veggies.
Any vegetable or 100% vegetable juice counts as a member of the Vegetable Group.
Fill half your plate with fruits and vegetables.
Fruits
Focus on fruits. Whole fruit is preferable to juice but any fruit counts: fresh, frozen, canned, 100% juice or dried.
Fill half your plate with fruits and vegetables.
Grains
Make at least half your grains whole.
Read labels to find more whole grain foods.
Whole wheat, oatmeal and brown rice are all good.
Protein
Go lean with protein.
Keep portion to 1/4 of the plate.
Nuts, beans/peas, seeds, poultry, lean meat, seafood, soy and eggs are in this group.
Dairy
Get your calcium-rich foods.
Remember to buy skim milk or 1% milk.
Go easy on cheese.
Skim yogurt is a good choice, too.

1

Lesson 3

How Fresh Is Your Food?

The Effects of Preservatives

To the Teacher

Different foods rot, or become moldy, at different rates. Mold is a fungus, like mushrooms, that grows from tiny spores in the air, unlike plants that grow from seeds. Generally, foods that are fresh will get moldy more quickly than foods that have been processed with preservatives because preservatives inhibit mold growth. While preservatives are good for ensuring that food that is transported long distances or remains on a store shelf doesn't rot quickly, it is not necessarily a good thing for our bodies. Fresh, organic foods with no preservatives may rot more quickly but are considered healthier for our bodies. In this long-term investigation, students will observe the rotting rates of various food items and learn about the effects of preservatives. The advantages and disadvantages of preservatives will be discussed. During this investigation, ingredients can be modified depending upon available resources.

Objectives

Students will make observations to understand the factors that influence mold growth. They will be introduced to the concepts of *preservatives, organic*, and *freshness*, and will evaluate the use of preservatives in food processing. **Safety: Do not use meats or fish as they will become rotted (and smelly) quickly; keep petri dishes or bags closed as mold can trigger allergies.**

Time Needed

One class period for setup: additional observation times in future classes

1

Materials

- 12 petri dishes or clear plastic sandwich bags
- Organic and conventional strawberries
- Other fruits such as grapes or bananas
- Processed fruit snacks (chews)
- A potato, cut into slices
- Potato chips
- A piece of cheese
- A processed, "cheesy" snack such as cheese puffs
- 1 loaf of bread from a deli
- 1 loaf of bread prepackaged (any brand)
- A tomato
- A can of tomato sauce
- A camera

Procedure

1. Show students a tomato and a can of tomato sauce. Using a Venn Diagram, ask students, "What do these two items have in common? How are they different?" Elicit responses (such as, "tomato is fresh; sauce is in a can; both have tomato in them; sauce is cooked" and so on). Ask, "What is the purpose of the can?" (Helps keep germs out of the sauce and the sauce usable for a long time.)
2. Explain that some foods are fresh while others are "processed." Processed means that they have been changed in some way, usually through cooking, canning, or using special chemicals called preservatives, to help them not spoil. Students will now be challenged to identify which foods may have preservatives through an experiment.
3. Students will now begin the long-term investigation by placing one of each food in a petri dish and labeling the contents. Photos should be taken of the foods so that comparisons can be made over time. Place dishes in a cabinet and for the next one to two weeks, have students observe and record how

these items change over time. In a science journal have students set up and make hypotheses about what they expect to occur during the long-term investigation. Have students record and draw observations daily for the "How Fresh Is Your Food?" experiment.

FIGURE 1.2.

Sample Setup for "How Fresh Is Your Food?"

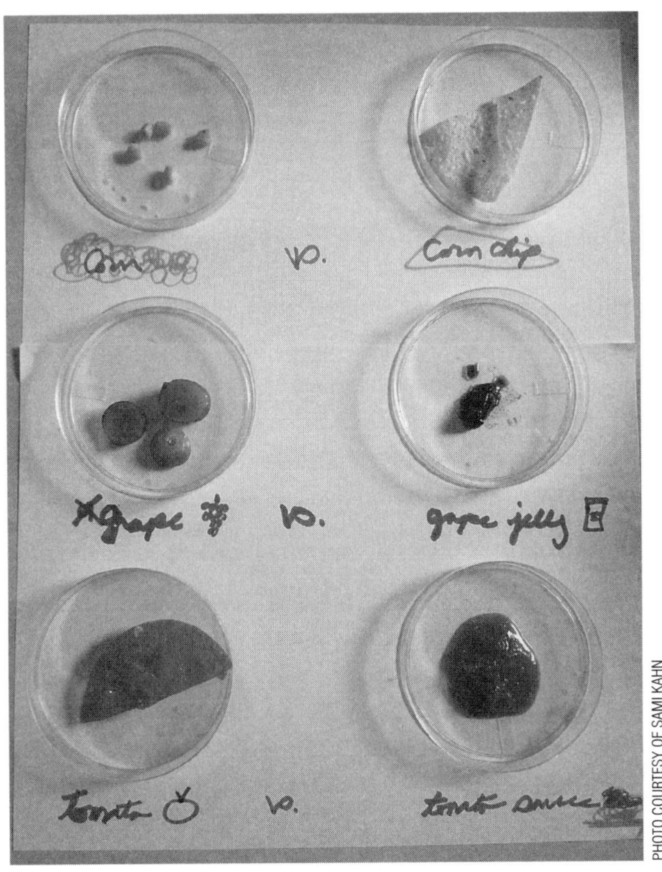

PHOTO COURTESY OF SAMI KAHN

Closure

1. Which foods rotted (developed fungus) more quickly? Why do you think this is so?
2. What are the advantages to preservatives? What are the disadvantages?

Assessment

Students will complete a "How Fresh Is Your Food" observation in their journal daily.

Scoring Rubric

1 pt.	2 pts.	3 pts.	4 pts.	5 pts.
Journal entries were attempted but all were incomplete.	Journal entries include sporadic observations and lack important details.	Journal entries were performed daily, but some entries are missing either written or drawn observations.	Journal entries include daily written and drawn observations, but observations lack some important detail.	Journal entries include daily written and drawn observations showing careful attention to details.

Accommodations for Students With Disabilities

Visual Impairments: Photos of the foods throughout their "rotting" progression can be taken and enlarged for observation; video projection system can also be used to enlarge images of the dishes; large magnifying lenses can be used for observations; a tactile template for the observations can be created to reinforce sections such as date, hypothesis, observations, and drawings.

Hearing Impairments: Consider using "Pass the Microphone" technique (any object can be designated as the "microphone") during whole-class discussions so that hearing-impaired students can determine who is speaking.

Learning Disabilities: Create sequence cards showing a photo of a piece of fruit progressing through the molding process; have students place cards in sequence to reinforce concepts; be sure to clarify the use of the word *fresh* as it can be used to refer to "unprocessed" foods, as we are doing in this lesson, or "preserved" foods, which means that it will not get moldy (but is sometimes referred to as "staying fresh)."

Motor/Orthopedic Impairments: For students with fine-motor skills challenges, use large, gallon-size plastic bags rather than petri dishes for easier manipulation; make sure that all activities are done at an appropriate height for students in wheelchairs; observations can be recorded on large paper pads using "chunky" crayons or markers.

Emotional Disabilities: Use random assignment (i.e., numbering off, picking out of a hat, and so on) to determine who prepares which dishes; students may become upset by the "wait time" involved in a long-term activity: remind them that things are happening even when we're not watching!

1

Lesson 4

The Six Nutrients Mystery Matching Game and Reading Labels Carousel

To the Teacher

Nutrition is the science of food and how it benefits the body. There are six main nutrients that are used to help our body grow, function, and heal: (1) Carbohydrates; (2) Proteins; (3) Fats; (4) Vitamins; (5) Minerals; and (6) Water. Each one has a special function:

- Carbohydrates give us energy.
- Proteins build muscle and help with growth and repair of cells.
- Fats keep us warm and provide long-term energy.
- Vitamins help regulate body functions (for example, Vitamin C supports the immune system; Vitamin A helps our eyes).
- Minerals help our bodies' systems to work (for example, calcium helps build bones and teeth, magnesium helps nerves to function).
- Water is used to carry all of the nutrients and wastes around our bodies.

Learning to read nutritional labels is an essential skill for making healthier choices. In particular, comparing similar foods for fat, protein, sugar, and fiber content can help to distinguish between alternatives. In this lesson, students will be introduced to food nutrients and label reading skills through an interactive computer game and will then visit "Label Stations" to practice their new skill.

Objective

Students will be able to identify the importance of each of the six nutrients and will be able to compare and contrast nutritional values of foods from their labels.

Time Needed

Two class periods

Materials

Props per team including water bottle (or photo of an H_2O molecule for older students), picture of a rocket, Legos, a scarf, a bone, and an empty vitamin bottle (or a photo of the Sun for Vitamin D or an orange for Vitamin C for older students), six cards (each with one of the nutrients listed on it), computer with projector, 6 food items with labels (or just the labels), chart for students (included below) to compete as they rotate.

Procedure

1. Show students a label from a common food, such as a cereal box or spaghetti sauce.
2. As a class, brainstorm some of the listed nutrients and discuss their purpose/function in the body.
3. In small groups, have the students try to match the prop to the nutrient name. (The answers are: Rocket = Carbohydrates (for energy); Sweater = Fats (for warmth); Bone = Minerals; Water Bottle or H_2O molecule = Water; Legos = Protein (for building blocks); and Vitamin Bottle (or sun or orange) = Vitamins
4. Introduce Nutritional Labels by playing "Ride the Label" at *www.nourishinteractive.com/kids/healthy-games/7-ride-the-food-label-game-nutrient-information* as a class.
5. Create "stations" around the room. Each station will have one labeled food item (or just the label).
6. As students rotate through each station in small groups, they will record nutritional data on the sheet provided.
7. Students will complete the questions at the bottom of the data sheet.

Closure

Ask questions such as: What are the six nutrients and how do they help us? Which foods had the most protein? Which foods contained the most sugar? Anything surprise you? Investigate the labels; why is it important to eat a variety of foods?

Assessments

Mix up props and cards and have students individually (or in pairs) match them again with explanations; hold up a prop and ask what nutrient it represents. Have older students create their own nutrient matching games with different props. Give students two new food labels to compare and ask them to evaluate the various nutrients contained in each.

Accommodations for Students With Disabilities

Visual Impairments: Use objects rather than photos for the props; if photos need to be used, make them very large or use "puffy paint" to make them tactile; create Braille matching cards or tactile cards using Wikki Stix; provide magnifiers for students with low vision in order to read labels more easily; create large-print labels by copying a food label and enlarge using a copy machine (you can do the same for the data sheet); allow student to use personal monitor for the online game so that he or she can enlarge print and/or zoom in; labels can be translated into Braille using a manual Braille label maker.

Hearing Impairments: Have students take turns at their tables picking up one item at a time: Only the student holding up an item should be speaking (allows the student to focus on one speaker at a time); put written instructions on the board to describe the process of moving around stations; use a visual cue (such as lights on/off) to signal when students are to move between stations.

Learning Disabilities: Keep a list (pictures and words) of nutrients on the board to reinforce the names, functions, and examples; have students match props to names and then names to props in order to reinforce the concepts; add visual cues (such as a picture of a muscle next to protein, a sugar cube next to sugar, and so on) to remind student what each nutrient represents; use paper or ruler to cover nutritional information that is not being compared (to help students focus on nutrient of interest).

Motor/Orthopedic Impairments: Velcro can be place on the cards and on a large sheet of poster board in order to keep them organized; Velcro can also be placed on a glove and on props so that a student with fine-motor skill challenges can lift up

props easily; be sure that this activity is done at a height that is comfortable for students in wheelchairs; use large items in boxes, cans, or plastic rather than glass for easier handling; modify the data sheet to allow for circling or marking of quantities rather than "write-ins."

Emotional Disabilities: Focus on turn-taking skills during the matching part of the activity: You can facilitate this by assigning students numbers in their teams (or having them pick number cards). If students disagree about the matching answers, have students take turns explaining their ideas; have other students "mirror" what the first student said to be sure that they are hearing and understanding each other; remind students that it is OK to disagree as long as they do it respectfully and have reasons for their answers; encourage groups to take turns "being the first label reader" at each station; consider having duplicate items at each station if groups are too large to easily share one item; praise students for waiting until the signal to change stations is given.

Name: ______________________________ Date: ______________

Six Nutrients Mystery Matching Game

Water	**Minerals**
Prop: Purpose:	Prop: Purpose:
Carbohydrates	**Fats**
Prop: Purpose:	Prop: Purpose:
Protein	**Vitamins**
Prop: Purpose:	Prop: Purpose:

Name: ______________________________ Date: ______________

"Read the Label" Stations Carousel

Station #	Food	Nutrition Information
1.		Fat – Calories – Sugar – Protein – Fiber –
2.		Fat – Calories – Sugar – Protein – Fiber –
3.		Fat – Calories – Sugar – Protein – Fiber –
4.		Fat – Calories – Sugar – Protein – Fiber –
5.		Fat – Calories – Sugar – Protein – Fiber –
6.		Fat – Calories – Sugar – Protein – Fiber –

1. Circle the food that had the most protein.
2. Put an "X" on the food that had the most fat.
3. Put check marks next to the two foods with the highest fiber.
4. Put a line through the food with the most sugar.

1

Lesson 5

Digestion

The Path to Good Nutrition

To the Teacher

Topic: Digestive System
Go to: *www.scilinks.org*
Code: SSI002

Digestion is the process through which the foods we eat get broken down into smaller molecules that can be absorbed and used by the body. Wastes that cannot be used are passed out of the body. There are two processes of digestion: (1) Mechanical Digestion: physical breaking down of food, and (2) Chemical Digestion: using enzymes or acids to break down foods. Absorption is the process by which nutrients and water are absorbed into the bloodstream. The path of food through the digestive system is:

1. **Mouth:** Mechanical digestion (chewing) and chemical digestion (saliva)
2. **Esophagus:** Muscular tube that carries food down from mouth to stomach – uses muscular contractions called *peristalsis.*
3. **Stomach:** Mechanical (contractions) and chemical digestion (stomach acids)
4. **Small Intestine:** Site of many enzymes that help with chemical digestion
5. **Large Intestine:** Absorption of nutrients and water into the bloodstream
6. **Rectum:** Muscular end of large intestine where solid wastes leave the body.

Many other organs, such as the liver, pancreas, and gall bladder, aid in the digestion of food by providing enzymes and other substances to help break food down into smaller molecules. In this lesson, students will learn about the digestive process and create "digestive vests" to follow the path of food through their digestive systems.

Objective

Students will be able to demonstrate the digestive process and identify key organs associated with digestion.

Time Needed

One class period

Materials

The Magic School Bus for Lunch video, paper bags, goldfish snacks, digestive cut outs and glue (or crayons for drawing organs if preferred), digestive diagram, nutrition journal

Procedure

1. Begin by biting into a cracker or an apple and chewing in front of students. Ask, "What happens after the food goes in my mouth?" Let's find out!
2. Watch *The Magic School Bus for Lunch* as an introduction.
3. Using the digestive system template and paper bags, have students make their own "digestive vests" by cutting out armholes and a hole for the head, and then gluing organ cutouts on (or drawing organs on).
4. Have students demonstrate eating a goldfish in order to model the digestive process.
5. Complete the digestion diagram in the journal.
6. Record and draw observations for "How Fresh Is Your Food?"

Closure

"What is the job of the digestive system?" (to break down food so that it can be used by our bodies) "How does food get used by our bodies?" (for energy, growth and repair)

Assessment

Have student teams coordinate a demonstration (using song, dance, drama) of food going through their digestive systems. Students will complete a digestion diagram in their journal.

Scoring Rubric

1 pt.	2 pts.	3 pts.	4 pts.	5 pts.
Team presentation demonstrates a lack of understanding of the path of food through the digestive tract and lacks creativity, collaboration, and effort.	Team presentation demonstrates a cursory understanding of the path of food trough the digestive path although team made reasonable effort on design of presentation.	Team presentation demonstrates general understanding of the path of food through the digestive tract but contains some inaccuracies.	Team presentation demonstrates accurate understanding of the path of food through the digestive tract, but lacks collaboration or creativity.	Team presentation demonstrates accurate understanding of the path of food through the digestive tract in a creative, collaborative manner.

Accommodations for Students With Disabilities

Visual Impairments: The digestive diagram can be made tactile by using Wikki Stix (wax covered strings); make digestive cutouts tactile using "puffy paint"; for a larger path of digestion, trace student's outline on butcher paper and create a life-size digestive tract.

Hearing Impairments: Use closed captioning on video; give student vocabulary in advance to clarify definitions; give all directions in both oral and written (or pictorial) formats.

Learning Disabilities: Practice sequencing before discussing the digestive system. For example, ask students to trace their path from waking up in the morning to getting to school; scaffold diagram by starting with only three to four organs to identify and then slowly increase by "uncovering" lines.

Motor/Orthopedic Impairments: Use adaptive scissors for cutting; use Velcro for attaching organs to vest; Velcro can also be used to make vest easier to put on by cutting open shoulders on bag and reattaching with Velcro so that the vest can be put on without hand raising.

Emotional Disabilities: When teams are deciding on their presentation, help students to compromise (if needed) by playing, "My Contribution." This is done by having each student give one favorite idea that will be used in the presentation. In this way, everyone makes a contribution but no one gets all of their ideas used. Of course, voting for favorite ideas can also be used if all agree to abide by the outcome.

Name: ______________________________ Date: ______________

Where Does My Food Go?

Digestive System Path

1. Label the organs.
2. Color the digestive track.

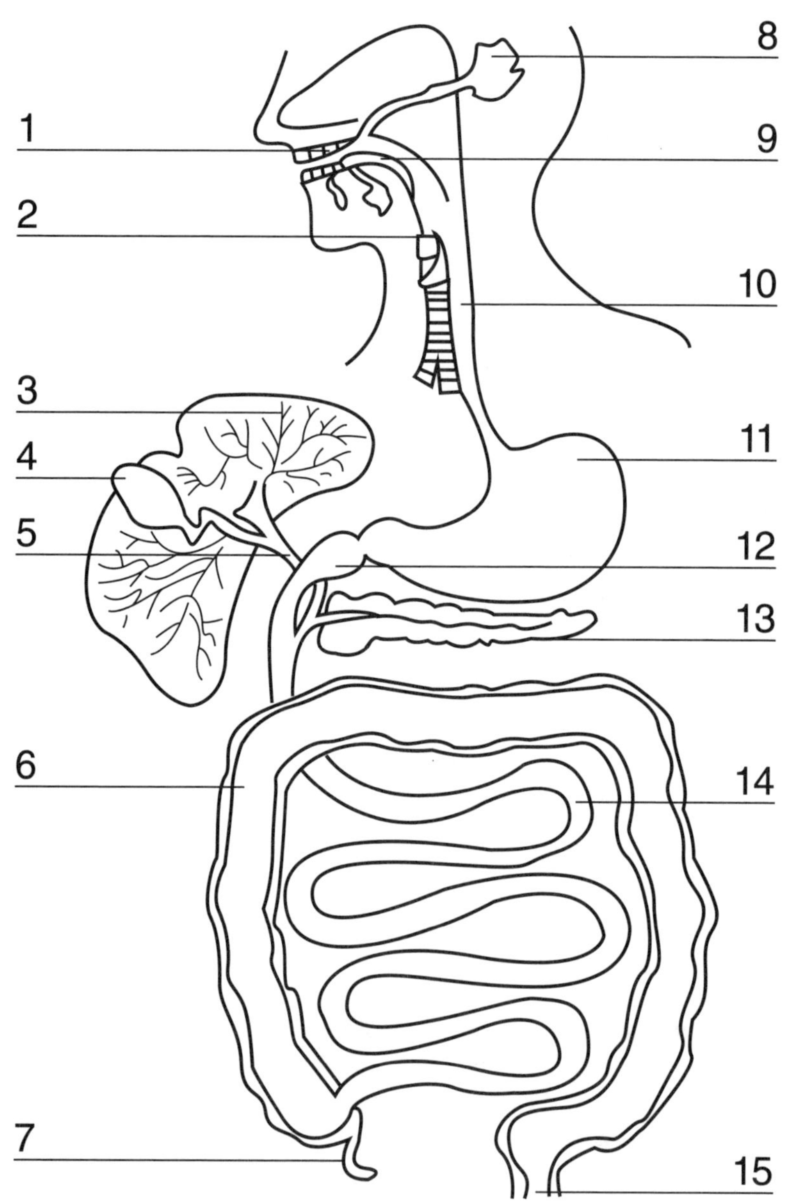

Answer Key

1. Mouth
2. Epiglottis
3. Liver
4. Gall Bladder
5. Bile Duct
6. Large Intestine
7. Appendix
8. Salivary Gland
9. Tongue
10. Esophagus
11. Stomach
12. Duodenum
13. Pancreas
14. Small Intestine
15. Rectum

Lesson 6

Find the Fat ... and Cheeseburgers on Trial!

Topic: Nutritional Disorders
Go to: *www.scilinks.org*
Code: SSI003

To the Teacher

Many popular foods contain large amounts of fat. Fat can be identified in food as it leaves a "grease" mark on paper bags, which absorb the fat. In this activity, students will conduct a paper bag grease test on various snack foods to identify whether the nutritional labels accurately depict the amount of fat and identify lower-fat alternatives. Students will then use readings provided to research the question, "Why do healthy foods matter?" and conduct a trial on whether cheeseburgers are "healthy" or "unhealthy."

Objective

Students will be able to distinguish between high- and low-fat foods by using a paper bag grease test. Students will identify healthy food choices and alternatives. Students will evaluate the nutritional value of a food with both high nutrition and low nutrition content to determine whether they feel it is "healthy."

Time Needed

Two class periods

Materials, Part 1

Fat Test: 1 paper bag per team (can cut bag into two halves so that it is a single layer); eight snack foods including: plain popcorn, buttered microwave popcorn, chocolate chip cookies, low-fat chocolate chip cookies, potato chips (fried), baked potato chips, pretzels, raisins; nutritional labels from each of the foods; stick of butter with nutritional label; bottle of cooking oil; a small amount of water; data sheets

Materials, Part 2

Printouts of the following articles:

- "How Eating Healthy and Unhealthy Foods Affects Your Body": *www.livestrong.com/article/41294-eating-unhealthy-foods-affects*
- "What Kids Say About What They Eat": *http://kidshealth.org/kid/talk/kidssay/poll_healthy_eating.html*
- "Why Should I Eat Healthy?": *www.livelifewell.nsw.gov.au/healthyeating*
- "Why Eat Healthy?": *http://www.fitness.gov/eat-healthy/why-is-it-important*
- Printout of McDonald's Nutritional Values: *www.nutrition.mcdonalds.com/nutritionexchange/nutritionfacts.pdf*

Procedure, Part 1

1. Show students a stick of butter. Ask them, "What is butter made of?" (cream). Read the fat content from the butter label. Explain that butter contains a large amount of fat. Show students a bottle of cooking oil and read the amount of fat contained in a serving.
2. Rub a small amount of butter on a paper bag and mark the spot, "butter." Rub a small amount of oil on the paper bag and mark the spot, "oil." Place a small amount of water on the bag and mark the spot, "water." Hold the bag up to the light. Students will notice that the butter and oil spots become translucent, but the water spot remains opaque. After a few minutes, the water spot will "disappear" (evaporate) but the butter and oil stains remain.
3. Explain that fats get absorbed into the paper and stay there, whereas the water evaporates. We can use the paper bags to test whether different snack foods have fat.
4. Have teams create eight boxes on their bags and label them with the food items listed in the materials section. Have students place the food items on the bags and allow them to sit several minutes. (If you want to continue this activity to a second period, cover the experiments with plastic wrap or place them in a large bag to prevent them from attracting insects.)
5. Have students record their results on the data sheet.

1

Closure

Which foods contained the most fat? (fried potato chips, buttered popcorn, regular chocolate chip cookies) Which foods were lower-fat/no fat alternatives? (baked potato chips, plain popcorn, low-fat cookies, raisins, pretzels). Share low/no fat foods with students (check for allergies).

Assessment

Students will compare their results to the labels on the foods. Did their results support the information on the label? Students interpret their findings and explain their results.

Accommodations for Students With Disabilities

Visual Disabilities: Distinguishing between "grease spots" may be difficult for a blind or low-vision student; for a low-vision student, shine a light through the paper bag to accentuate the spot; allow students to touch the spots and feel for the grease.

Hearing Impairments: Provide all instructions in multiple formats, including verbal and written (or pictorial).

Learning Disabilities: Help students to organize their experiment by demonstrating ways to create eight boxes on their paper bag (folding, using a template, or creating "tic-tac-toe" type designs); be sure that students label their boxes before receiving the food items; use highlighter on food labels to highlight fat content.

Motor/Orthopedic Impairments: Be sure that all demonstrations and experiments are performed at a level (height) that is comfortable for the student; use chunky crayons for box creation/labels, or create sticky labels for student to place on bags; increase size of data sheet.

Emotional Disabilities: Reiterate rules about eating in the science lab before food is handed out; help teammates to divide team roles either by food number (each student in a team of four is responsible for setup of two foods), or by jobs (one student is materials manager, one is setup person, one is principal investigator, one is presenter, and so on).

Procedure, Part 2

1. Divide the students into small groups.
2. Give each group copies of one article. Each group will become experts on its article.
3. As the groups read and discuss the articles, have them create a list of the benefits of healthy eating and some of the problems that may arise from unhealthy eating by answering the question, "Why do healthy foods matter?"
4. Discuss results as a class.
5. Divide the class in half and show them the nutritional content of a fast food restaurant cheeseburger.
6. Assign half the class to argue "for" the cheeseburger as a "healthy" food, and half the class to argue "against" the cheeseburger as a healthy food. After students argue, vote for a verdict.

Closure

Ask, "Why does eating healthy food matter? How can some foods be healthy and unhealthy at the same time? What are some ways we could make a "mixed" food more healthy?"

Assessment

Students can be assessed on their participation in group discussions about the articles and during the trial.

Extension

Have groups revisit the grease test results from the last class. After looking at the nutrition labels for those items, students should draw a picture of one of the items and explain why they feel it is or isn't healthy.

Accommodations for Students With Disabilities

Visual Impairments: Prepare Braille or large-print handouts of readings.

Hearing Impairments: Use "round robin" technique for reading so that it is obvious who is reading and who reads next; use "pass the microphone" technique during the trial to ensure that only one student speaks at a time.

Learning Disabilities: Provide students with graphic organizers to sort facts; allow students to utilize highlighter pens to highlight important facts.

Motor/Orthopedic Impairments: Allow student to record answers to questions if writing is not possible; Clear pathways between student's home team and trial areas.

Emotional Disabilities: Some students may have difficulty being assigned to a trial group representing opinions with which they don't agree. Remind students that they are only focusing on that opinion in order to get information—they will be given a chance to voice their own opinions at the end of the activity.

Name: ______________________________ Date: ______________

Find the Fat

Box	Food Item	Observation
Box 1		
Box 2		
Box 3		
Box 4		
Box 5		
Box 6		
Box 7		
Box 8		

Lesson 7

Kids as Consumers

Evaluating and Creating Food Commercials

To the Teacher

Armed with some knowledge about good nutrition and an understanding of the ways in which advertisers often try to market unhealthy foods to children, students will have an opportunity to evaluate commercials and then create their own commercial for a healthy food aimed at children.

Objective

Students will be able to evaluate the aspects of commercials that target children, create a commercial for a healthy food item that is appealing to children, and identify the healthy aspects of the food.

Time Needed

Two class periods

Resources

computer, video (*www.youtube.com/watch?v=5Jlv1c-3JeM*), props for commercials, flip video recorder.

Procedure

1. Teacher will show students a video that illustrates multiple food commercials aimed at children *www.youtube.com/watch?v=5Jlv1c-3JeM*.
2. After viewing the video, have small groups discuss their opinions of the commercials, which targeted unhealthy foods to children. Next discuss as

a whole class. Do they think it is ethical for advertisers to target children? (Some may say that advertisers should be able to do whatever they want if people will buy the product; others may find it ethically objectionable.) Discuss the idea that it is important for consumers to be knowledgeable about nutrition so that they can make good choices regardless of the advertising.

3. Students will create their own 30-second commercial aimed at children for a healthy food item of their choice.
4. Students will record their video using a flip video recorder.
5. Students will play their commercials for the class. As students watch the videos, they will record whether the commercial was appealing to them and if they would buy and eat the product in the commercial in their nutrition journal.

Closure

Ask, "How did it feel to create a commercial? What techniques did you use to make your commercial appealing?"

Assessment

Students will be assessed by their commercial using a rubric that includes whether students identified whether the food item was healthy, why is it healthy (with three or more reasons), and if the commercial is appealing to children.

Extension

Students can identify other products that use strategies similar to those used to sell foods aimed at children (such as toys, vacation/recreation spots, clothes, and so on).

Accommodations for Students With Disabilities

Visual Impairments: Show commercials on large screen; send link for commercials home to parents so that they can watch and describe to student in advance; have students use descriptive language when discussing commercials; require students to have dialogue in commercials and print dialogue in Braille if needed.

Hearing Impairments: Use closed captioning on commercials; have students include captions in their commercials or print a script.

Learning Disabilities: Review instructions for camera and computer program usage in both verbal and pictorial forms; provide students with a checklist rubric of the components of a good commercial.

Motor/Orthopedic Impairment: A small tripod can be used for camera so that student can locate the best position for the camera and set it down; Be sure that there is ample space for student to move around for commercial rehearsal.

Emotional Disability: Emphasize cooperative skills during commercial-making process; remind students that brainstorming techniques allow all ideas to be recorded; help students come to consensus on ideas by encouraging voting, using "one idea" from each teammate, and compromising (perhaps a student whose idea for the commercial isn't used can be in charge of directing the commercial).

Name: ______________________ Date: ____________

What Did You Think About This Commercial?

Commercial	Would you eat this item? Why or why not?	Is this commercial appealing to children? Why or why not?

Scoring Rubric

Student(s) Name(s):____________________ Commercial:______________________

Commercial Rubric

Did the commercial state whether the food item was a healthy food item?	____ 5 pts
Did the commercial state why the food item was healthy, giving three or more reasons?	____ 15 pts
Is this commercial appealing to children?	____ 10 pts
Did the members of the group participate equally?	____ 5 pts
Was the commercial creative?	____ 5 pts

Total Pts.____________/ 40

1

Lesson 8

Final Debate and Letter Writing

Should Schools Charge More Money for "Unhealthy" Foods?

To the Teacher

At the start of this unit, students were faced with the question of whether schools should charge more money for unhealthy foods. Votes were cast and explanations were given. Over the last several lessons, students have learned about the importance of healthy eating, explored strategies for making better choices, and become aware of the way advertisers appeal to children. In this lesson, students will once again cast votes (providing their reasons using their new knowledge), revisit their original "fantasy lunch" from the first lesson, and write a letter to the school principal explaining their thoughts.

Objective

Students will once again vote on the question, "Should schools charge more money for unhealthy lunches?" and write a letter to the principal explaining their thoughts on the idea of increased prices for unhealthy food in the cafeteria.

Time Needed

One to two class periods

Materials

Voting materials used in Lesson 1, current school lunch menu

Procedure

1. In a whole-class discussion, elicit ideas from students about the facts they have learned about nutrition (food groups, fat content, nutrients, digestive system, impact of eating choices on health, ways in which advertisers target children, and so on).
2. Remind students that on the first day of the unit, they voted on whether schools should charge more money for unhealthy foods, an idea called a "fat tax." Review the results from the first class.
3. Repeat the vote. Allow students to discuss their ideas, reminding them that there is no wrong opinion here. (Some students might suggest that it is OK for schools to charge more for unhealthy foods because it is up to the student to make good choices anyway; others may say that it is a good idea to prevent students from buying them; still others may say that the price increase won't change anyone's mind; and some may suggest that it is not fair for the school to "control" their eating this way. These are just some of the possible arguments.) What is important is that students are able to articulate a rationale for their opinion and use some of the scientific facts that they have learned in their discussion.
4. Have the students write a letter to their principal explaining their thoughts on charging more money for unhealthy school lunch items. The letter should include at least five facts they've learned from this unit, and should demonstrate that they have a sound knowledge of proper nutrition. Students should also recommend a sample lunch menu that is healthy and contains well-liked food items.

Closure

Share some of the letters with the class. This can be done with students identified or anonymously. Share students' original "fantasy" meals from the first lesson. What do they think of them now?

Assessment

The letter to the principal should include at least five facts they have learned throughout this unit and will be graded according to the attached rubric. The sample lunch menu should satisfy the "My Plate" requirements as discussed in Lesson 2.

Accommodations for Students With Disabilities

Visual Impairments: Allow student to write letter on a computer with large display if this is most comfortable for the student; speech-to-text software is also available to assist with this; see Lesson 3 for modifications for voting.

Hearing Impairments: Allow student to read copies of letters while students are presenting them; write student responses to questions on the board or on sentence strips for clarity; see Lesson 3 for modifications for voting.

Learning Disabilities: Provide a letter-writing template and/or graphic organizer to help student organize letter; use sentence strips to write down facts that are brainstormed during initial discussion so that they can be referenced during letter-writing.

Motor/Orthopedic Impairments: Use pencil grips and paper guides for students with fine-motor skills challenges; incorporate assistive technologies such as voice-to-text programs for students who do not have use of their hands.

Emotional Disabilities: Continuously remind students that there is no "right answer" to the "fat tax" question; encourage students to focus on facts that they have learned rather than simply getting others to agree.

Scoring Rubric For Letter

Student Name: ____

Did the letter state the student's opinion on the unhealthy food price increase?	____ 5 pts
Did the letter provide at least five specific facts learned from this unit?	____ 15 pts
Does the letter include a possible lunch menu?	____ 10 pts
Does the menu meet the "My Plate" requirements?	____ 5 pts
Does the letter contain proper grammar and structure?	____ 5 pts

Total Pts.____________/ 40

Optional Lesson

Field Trip to Supermarket and "Eat This, Not That" Slide Show

To the Teacher

One of the most important aspects of nutrition that students can learn is to make good choices and identify healthier alternatives to the unhealthy foods that they might like. Now that students have become aware of marketing tactics aimed at children, they can determine which choices are good and which are not and learn to identify substitutes for the latter. In this lesson students will visit a supermarket, take photos of foods that they feel are marketed to them, and take photos of healthier alternatives. Students will then create an "Eat This, Not That" presentation.

Objective

Students will be able to identify foods that are marketed specifically toward them and determine whether they are healthy or unhealthy. For unhealthy options, they will be able to identify a healthier alternative.

Time Needed

Two class periods

Materials

Digital cameras, uploading cords, computer, adult chaperones (If available, the book *Eat This, Not That*, by David Zinczenko and Matt Goulding, can be displayed.)

Procedure

1. Show students an example of a product (such as a cereal box) that is clearly aimed at children. Ask them to review some of the methods companies use

to target children in advertising (bright colors, toys in the boxes or contests, cartoons, famous characters, and so on).

2. Explain to students that they will be taking a field trip that is a special kind of scavenger hunt: they will be searching for products aimed at them as well as some healthier alternatives.
3. Take students to a supermarket. They will be working in small groups, with an adult chaperone in each group.
4. Using digital cameras, they will take pictures of foods that are specifically marketed toward kids.
5. They will also take pictures of the nutritional information for each food they find.
6. For any food they determine to be unhealthy, they will look around the store and find a healthier alternative. They will take a picture of this product as well as its nutritional information.
7. Back at school, students will upload their pictures to the computer.
8. Using the pictures, they will create a slide show "book" of "Eat This, Not That." For example, the first slide may contain a picture of graham crackers with its nutritional information, saying "Eat This," and the next slide, entitled "Not That" would have a picture of icing-covered animal crackers with sprinkles.

Closure

Ask, "What did you notice about the foods targeting children? Was it difficult to find healthier alternatives? Did you notice anything about the placement of food targeting children in the supermarket?"

Assessment

1. Student teams should have at least 10 slides in their slide show.
2. Student teams should have at least 20 pictures from the supermarket.
3. Students will complete their observation journal for How Fresh Is Your Food?

Accommodations for Students With Disabilities

Visual Disabilities: During the trip to the supermarket, encourage teammates to describe the details of the items they see; add narration, either recorded or live, for the slide show.

Hearing Disabilities: Give specific instructions on the boundaries of the supermarket (such as, "We are focusing on aisles 1–5,") to prevent wandering; use captions on slide show.

Learning Disabilities: Bring pictorial examples of products targeting children on field trip to remind students of the cues; focus on key words, such as "fruit" so children can find a child-targeted food, such as "Fruit Loops" and then a healthy fruit.

Motor/Orthopedic Impairment: Make sure that the supermarket selected is easily accessible in terms of entryway, sufficiently wide aisles, and of course, consider the mode of travel to the supermarket. Discuss with student their ideas about the level of support needed on the field trip and discuss final plans with parents.

Emotional Disabilities: Discuss the procedure for sharing cameras among teammates before leaving for trip; allow students to give input on how this will occur (such as the camera gets passed after every shot or after every two shots); discuss expectations for behavior and consequences before leaving for trip; if inappropriate behavior occurs at supermarket, designate a "time out" area outside the store with an adult chaperone so that the student will experience a consequence of inappropriate behavior until he or she settles down. Be sure to communicate the reason for the "time out" and the future positive expectation before returning to the store.

Note: "Eat This, Not That" refers to the book *Eat This, Not That* by David Zinczenko and Matt Goulding. New York: Rodale, 2008.

Unit 2

Animals at Work

Should animals perform in circuses?

Unit Overview

In this unit, students will engage in several activities and investigations about working animals in order to gain relevant science content and process skills, as well as an appreciation of working animals. Students will then debate the question of whether animals should perform in circuses, thereby modeling the research, reasoning, and discourse skills that are necessary for informed citizenship.

Topic: Animals in Circuses
Go to: *www.scilinks.org*
Code: SSI004

Key Science Concepts

Needs of Living Things, Classification, Animal Senses, Forces, Applications of Technology

Ethical Issues

Animal Rights and Welfare, Stewardship, Ethics and Economics in Entertainment

Science Skills

Classifying, Observing, Measuring, Recording and Interpreting Data, Understanding Cause and Effect, Communicating Results, Forming Arguments From Evidence

Grade Level

Elementary

Time Needed

This unit includes 10 flexible lessons that can be adjusted to the pace and schedule of your class. Note that Lessons 2–5 can be done in any order.

Lesson Sequence

- **Lesson 1.** All Sorts of Working Animals (Classification/Argumentation/NOS)
- **Lesson 2.** Putting the Horse Before the Cart (Draft Animals; Pulling Forces)
- **Lesson 3.** Follow Your Nose (Tracking Animals Using Senses)
- **Lesson 4.** Dolphins With a "Porpoise" (Using Echolocation for Rescue)
- **Lesson 5.** Meet SomeAnimal Helpers (Therapy and Service Animals; Bios and Learned Behaviors)
- **Lesson 6.** How Do You Feel About Working Animals? (Values Clarification)
- **Lesson 7.** Circus Animals (Investigating Diets and Habitats)
- **Lesson 8.** Under the Big Top: Animals in the Circus Debate (Research and Role-Play)
- **Lesson 9.** Where Do You Stand? (Argumentation/Dialogue)
- **Lesson 10.** Culminating Project (Whole Class e-book on Working Animals)

Background on the Issue

The history of humans' relationship with animals has been one filled with stories of tremendous reverence and friendship as well as cruelty and abuse. The human "right" of dominion over animals has compelled people to use animals for much more than just food or clothing. Evidence of "beasts of burden," or working animals, appears from the earliest human records, with cave records of early humans using dogs in hunting.

Working animals are those that are usually domesticated and are trained to perform tasks. Some of the categories of working animals include:

- Draught (or Draft) Animals: Used for their physical strength
- Riding (or Mount) Animals: Selected for speed and endurance
- Pack Animals: Used for carrying loads

- Sensory Animals: Capitalize on unique senses or behaviors—search and rescue, hunting, herding, human assistance animals, finding plants, guarding
- Performing Animals: Used for entertainment and sport

The use of animals in circuses has been particularly controversial, as there have been some highly publicized incidences of abuse and neglect. Circus advocates argue that the animals are well cared for and that they serve important roles in entertainment, education, and tradition. The animals also provide employment opportunities for circus employees. Animal rights advocates argue that circuses are inhumane, and that using animals for entertainment is frivolous and unnecessary.

In deciding whether animals should perform in circuses, students touch on a fundamental topic taught in every elementary science course: What do animals need to live? The typical science course response is, "Animals need food, water, air, light, and space/shelter to survive." But the question posed in this unit also considers a "quality of life" component, requiring both a moral and biological assessment. Because of this, the issue lends itself perfectly to SSI.

Connecting to *NGSS*

K-PS2-1. Plan and conduct an investigation to compare the effects of different strengths or different directions of pushes and pulls on the motion of an object.

K-LS1-1. Use observations to describe patterns of what plants and animals (including humans) need to survive.

1-LS1-1. Use materials to design a solution to a human problem by mimicking how plants and/or animals use their external parts to help them survive, grow, and meet their needs.

3-PS2-1. Plan and conduct an investigation to provide evidence of the effects of balanced and unbalanced forces on the motion of an object.

3-LS3-2. Use evidence to support the explanation that traits can be influenced by the environment.

4-LS1-1. Construct an argument that plants and animals have internal and external structures that function to support survival, growth, behavior, and reproduction.

4-LS1-2. Use a model to describe that animals' receive different types of information through their senses, process the information in their brain, and respond to the information in different ways.

Accommodations for Students With Disabilities

Visual Impairments: Each lesson in this unit provides specific instructions for making materials tactile; be sure to describe all photos and illustrations verbally.

Hearing Impairments: The activities in this unit will undoubtedly prompt excitement and noise. Monitor the sound level in the room during group discussions as background noise can negatively impact students with hearing loss as well as interpreters.

Motor/Orthopedic Impairments: Several of the activities in this unit require movement around the classroom. Make sure that aisles are wide and all materials are easily accessible. If a student is unable to stand, have his or her group sit during group presentations so that all group members are at the same level.

Learning Disabilities: This unit involves a variety of lessons with several different "big ideas." An O-W-L chart (observed, want to know, learned) will help students to keep their learning organized and serve as reinforcement of ideas.

Emotional Disabilities: The activity level of this unit will engage students, but may also be overstimulating for some. Allowing students to take a voluntary "time out" as needed can be helpful, as can a clear set of behavioral goals for each lesson.

This unit is based on projects submitted by Sami Kahn, Andrea Churco, Lisa Clautti, and Katie Frost.

2

Lesson 1

All Sorts of Working Animals

To the Teacher

For thousands of years, humans have trained animals to do work for them. Some of the types of work include carrying, lifting, moving, pulling, plowing, tracking, guiding, hunting, and entertaining. In this activity, students are introduced to the many ways in which animals and humans have partnered to work together. Students will sort photos of working animals into categories and discuss their criteria for the categories.

Objectives

Students will be able to classify working animals into categories including draught, sensory, riding, pack, and performing.

Time Needed

One 50-minute class period

Materials

Book, *Animals at Work*, by Liz Palika and Dr. Katherine A. Miller (Howell Book House, 2009), photos of working animals per team, sorting mats or blank paper for labeling categories.

Photo ideas: Guide dog, herding dog, horse pulling carriage, dogs pulling sled, camel with rider, elephant lifting wood, dolphin performing or with radio transmitter, therapy cat, police dog, tiger/lion performing in circus, and so on.

Procedure

1. Ask your students, "What is work?" ("getting things done," "doing hard things," making an effort," "a job"). Suggest to students that in science, *work*

usually refers to using force to make something move. But when we think about work every day, it can also mean making an effort to do a job.

2. Ask, "Are people the only living things that do work?" (Students may discuss animals such as bees that make hives and honey, gophers that dig tunnels for nests, and animals hunting for food.) Suggest that sometimes people train animals to do work for them. Ask, "Can you think of any animals that work for humans?"
3. Show students the cover of the book, *Animals at Work*. It shows a guide dog helping a person using a wheelchair. Ask, "What is this dog doing?" "How does the dog know to guide the person?" "Could any animal do this work?" "Let's learn more about working animals."
4. Read selected pages of the book, showing a great variety of different types of work including pulling, pushing, lifting, carrying, herding, hunting, tracking, providing care and companionship, entertaining, and so on.
5. Give teams of students sets of photos of working animals. Challenge teams to determine a way to classify, or group, the photos so that they make sense to them.
6. After the team has decided on their groupings, have teams report to each other. Discuss the ways in which different groups may have chosen to classify the animals differently (i.e. perhaps by animal species rather than type of work, or by the way it moves (wings, legs, flippers).

Accommodations for Students With Disabilities

Visual Impairments: Prepare Braille cards with descriptions of animals, or utilize Wikki Stixx to make photos tactile; have partners describe the photo to student; provide a tactile sorting outline, such as string taped down to table to create sections.

Hearing Impairments: Use "Pass the Microphone" technique in order to ensure that only one student is speaking at a time in the team.

Motor/Orthopedic Impairments: Attach Velcro to photos and attachments on a sorting board so that the photos will stick and stay easily.

Learning Disabilities: Provide a demonstration of sorting items such as foods, transportation, clothes, and so on.

Emotional Disabilities: Emphasize the concept that all groupings are acceptable as long as students can give valid reasons for their choices; be aware that the "consensus" aspect of the team work may be challenging. Encourage students to express their opinion but also acknowledge those of others.

Closure

1. What categories did you use to classify, or group, your working animals?
2. Is it possible to have more than one correct way to classify the animal?
3. Do scientists ever disagree on how to classify things? If so, how do they decide on the "answer?"

Assessment

Teams present their groupings to the class. Groupings are memorialized on a sorting card with the categories and student names included.

Name: ______________________________ Date: ______________

Throughout the unit, add and take away items from this T-chart. List the pros and cons of using animals to work for people.

Pros ☺	**Cons** ☹

2

Lesson 2

Putting the Horse Before the Cart

To the Teacher

Many animals are used for pulling heavy loads. They are called "draught" or "draft" animals. Oxen, horses, elephants, mules, and sled dogs are some examples of strong animals used for pulling various things from carts to plows to sleds. What factors impact how hard (how much force) it is to pull a load? In this lesson, students will be introduced to the true story of "Blind Tom," a blind horse who helped build the transcontinental railroad by hauling heavy carts of materials. They will then test various loads with a spring scale to determine what factors make the work of moving loads easier or more difficult.

Objectives

Students will be able to identify factors that make pulling a load more difficult, including weight of the load, steepness of the incline, and surface condition.

Time Needed

Two 50-minute class periods

Materials

Book, *Blind Tom: The Horse Who Helped Build the Great Railroad,* by Shirley Raye Redmond, a spring scale per team, a toy horse (small plastic is fine or paper cutout), a wooden ramp or large piece of cardboard per team (along with a block or book to create an incline), a "load" to be pulled (can be a book, a bag of rocks or pennies, or gram weights), meterstick, and a data sheet for recording. Optional materials: sandpaper, aluminum foil, a toy car or cart, pencils.

Teacher Preparation

Attach the loads to the spring scales and place at each table along with ramps or cardboard pieces.

Procedure

1. Review concepts about working animals from last session. Ask, "What are some of the jobs that animals do for people?" (lifting, carrying, tracking, pulling). Show students a series of photos of draught animals. "What job are these animals doing?" "Why do you think these animals were chosen for this job?" (strong and high endurance)

2. Introduce students to "Blind Tom," the horse on the cover of the book. Ask students to anticipate how a horse could help build a railroad. Would being blind affect his ability to work?

3. Read, *Blind Tom: The Horse Who Helped Build the Great Railroad.* Discuss the role Tom had in building the railroad. Ask students, "Do you think every load Tom pulled felt the same?" "What do you think makes a load easier or harder to pull?"

4. Explain to students that their horses (either plastic or paper) are going to pull a load for one meter. We are going to measure how "hard" it is to pull the load by using a spring scale, which measures "force." Demonstrate attaching the horse to the front of the spring scale, and then attaching the load to the bottom of the scale. Pull the horse across the table and show students that the scale shows a measurement.

5. Students will then test the amount of force needed for the horse to pull the load one meter when the ramp is flat, and at various increments of steepness using blocks or books. Finally, see how much force is needed to lift the load straight up.

6. Allow teams to "double" their loads by combining loads between teams. Have students record their observations in lab notebooks.

7. If time allows, have students test the effects of different surfaces on the force. Covering the ramp with sandpaper and then aluminum foil will introduce concepts of friction as a force that slows movement.

Accommodations for Students With Disabilities

Visual Impairments: Spring scales can be marked using glue dots to identify markings; use spring scales with large, simple readouts; a Braille meter stick can be used on the ramp; a toy horse is preferable over paper for a student with a visual disability.

Hearing Impairments: Provide pictorial instructions for use of spring scale; allow student to pre-read the book with parents in advance of class to ensure that classroom noise doesn't impeded understanding if not signed.

Motor/Orthopedic Impairments: Use a plastic toy horse large enough for student to grasp easily; place materials at proper height if student uses a wheelchair. With a ramp, the best place might be the floor so that the student can look down at the scale.

Learning Disabilities: Review vocabulary prior to class; allow student practice time with spring scales weighing various objects.

Emotional Disabilities: Assist students in rotating roles throughout activity. Give students number cards so that the jobs can be easily rotated through each trial.

Closure

1. What are some of the things that affect how hard (how much force is needed) it is to pull something? (weight of the load, steepness of the incline, texture of the surface)
2. Why do we put wheels under carts and other heavy loads? Test this out by laying pencils down under the load and pull. ("wheels" reduce friction)

Language Arts Extension

What does the phrase, "Don't put the cart before the horse" mean? (idiom suggesting that things are done in a proper order). Name other idioms that use animals in them. (i.e., "it's raining cats and dogs," "the straw that broke the camel's back," and so on).

Assessment

Give teams different loads to test. Have students predict and explain findings.

Lesson 3

Follow Your Nose

To the Teacher

Many animals rely on their sense of smell for survival. Ants, for example, leave a scent trail marking food sources for other ants; the better the food source, the more scent markings are left on the trail. To define an area or territory, many animals mark objects with scent from special glands. Scent-marking can be used for finding a mate or for establishing an area for family, shelter, and food supply. Humans have unique personal scents which can be used to locate them when missing. Animals with acute senses of smell, such as dogs, are used for tracking to locate missing people or items. In this activity, students will simulate the experience of a tracking animal to locate a missing toy. They will explore their sense of smell and discover why smell is important to animals, including themselves. They will also practice measuring and mapping skills.

Objective

Students will be able to describe how the sense of smell is useful for survival for many animals, including humans.

Time Needed

One 50-minute class period

Materials

3 small bottles of flavored extracts (i.e., peppermint, vanilla, lemon), a doll, a piece of cloth, index cards, cotton balls, meterstick, paper.

Teacher Preparation

Scent-mark index cards by dabbing extract on a cotton ball. Mark 30 cards, 10 in each of three scents. Hide a doll somewhere in the classroom. Choose one of the scents to be the "tracking scent" (the one that will lead to the missing doll). Place the cards in a path (the windier the better) from one point in the room to the other where the doll is hidden. Place the other scented cards around the room as "decoys." Place a dab of the "tracking scent" on a piece of cloth. This will be given to student "hounds" to smell and try to follow that scent.

Procedure

1. Ask your students how their sense of smell is important to them. Did they ever have a cold and lose their sense of smell? How did it feel? Did it affect their sense of taste? Could your sense of smell save you from a dangerous situation? (Yes, you could detect gas, smoke, or rotten food.)
2. Next, ask students how different animals rely on their sense of smell. What purpose does smell serve for these animals? (It helps them find food, detect danger, find a mate, and identify another animal's territory.)
3. Show students photos of tracking dogs that are used for their sense of smell. Tell students that they are being asked to locate a missing doll based on the scent. Students will work in teams to follow the trail, and record on paper which marked cards led them to the doll
4. Challenge the teams to use their noses to find the trail to the missing doll. If they find a card that all members agree is scent-marked with the proper scent, they should map that card on their paper.
5. When the doll has been located, have the students walk the scent trail together. As a class, measure the distances between the cards on the doll scent trail and record measurements on maps.

Accommodations for Students With Disabilities

Visual Impairments: Have teammate guide visually impaired student to cards; mark each of the cards at approximately the same height; use Braille metersticks or a trundle wheel for measurement.

Hearing Impairments: Mark the perimeters of the study area clearly to prevent student from wandering out of visual contact.

Motor/Orthopedic Impairments: Assess the accessibility of the study site before performing the activity; if student is in wheelchair, be sure to scent-mark the cards at a height accessible to student.

Learning Disabilities: No modifications necessary.

Emotional Disabilities: Consider having teams take turns at finding scent-marked cards to limit noise and activity levels.

Closure

1. Why is the sense of smell important to animals? (Find food, locate offspring and mates, define territory, communication, or avoid danger.)
2. Why are certain animals used for tracking? (Very keen sense of smell, trainable, natural instincts to follow scents.)
3. Do people mark their territory? (Think about fences, hedges, "Private Property" signs, and so on.)
4. Do certain scents remind you of a place or person? Why do you think it's important for us to recognize familiar scents?

Assessment

Teams research how animals such as ferrets, rabbits, deer, cats, and dogs scent-mark their territories and present their findings to the class.

Scoring Rubric

1 pt.	2 pts.	3 pts.	4 pts.	5 pts.
Teams conduct limited, inaccurate research.	Teams conduct thorough research that contained inaccuracies.	Teams conduct accurate research that is presented in a clear manner, yet some important information was missing.	Teams conduct accurate and thorough research; however, the findings are presented in an unclear or unpersuasive manner.	Teams conduct accurate, thorough research and present their findings in a clear and compelling manner.

Lesson 4

Dolphins With a "Porpoise"

To the Teacher

Many working animals are used for their extraordinary senses. As was seen in the prior activity, dogs' amazing sense of smell is used to track humans and materials. Dogs' excellent hearing also makes them good guard animals. Dolphins have an unusual ability to use sound to locate items in water, even if it is very murky. Using *echolocation,* dolphins emit sound waves that bounce off surrounding objects and return to the dolphin, allowing it to get a "picture" of the area. Humans have made use of this skill by training dolphins to locate underwater mines in the military. The dolphins are also trained to rescue naval swimmers. In this activity, students will test out their own "echolocation" abilities by using blindfolds and tennis balls to identify their position.

Objective

Students will be able to collect and analyze data in order to draw conclusions about sound and echolocation.

Time Needed

One 50-minute class period

Materials

The video "Dolphin Life Savers: Training for a Mission" (*www.youtube.com/watch?v=8JNrtFG4RU4*), meterstick, masking tape, one blindfold per team, one tennis ball per team.

Procedure

1. Review prior class's discussion on the use of a dog's sense of smell to locate people or items. Ask, "Can an animal's sense of hearing be useful to

people, too? How?" Guard dogs, for example, are trained to bark when they hear intruders. Show student a photo of a dolphin. Elicit information from students about dolphins (mammals, related to whales and porpoises, use echolocation sonar). Tell students that dolphins actually work in the Navy. Doing what?

2. Explain echolocation to students. The dolphin can create clicking sounds that create sound waves that hit objects and bounce back to the dolphin, allowing it to know where the objects are, even without seeing them.
3. Show the video "Dolphin Life Savers: Training for a Mission," which describes the way dolphins are trained to locate underwater mines.
4. Students will then test their own "echolocation" abilities by simulating echolocation with a tennis ball (representing sound waves).
5. Have each team measure three intervals from a wall: 1 meter, 3 meters, 5 meters, and mark each interval with masking tape.
6. Allow students to practice rolling the ball (a sound wave) toward the wall and counting how many seconds it takes for it to hit the wall. Encourage the students to try to roll the ball with consistent force. They should find that it takes longer when they are standing farther from the wall.
7. Students will take turns being the "dolphin" by placing a blindfold on their eyes and allowing their teammates to guide them to one of the marked distances.
8. The "dolphin" then rolls the ball toward the wall and counts the number of seconds it takes until they hear the ball hit the wall. The "dolphin" then guesses how many meters he is from the wall. Students record data in lab notebooks.

Accommodations for Students With Disabilities

Visual Impairments: Using a string, tie knots in it at the 1-, 3-, and 5-meter points. Attach it to the wall and allow the visually impaired student to use the string as a measurement guide for leading other dolphins.

Hearing Impairments: Have a partner tap the hearing impaired student when the ball hits the wall; be sure to discuss the use of a blindfold with the hearing-impaired student in advance of the lesson, as the temporary loss of sight may be uncomfortable for the student.

Motor/Orthopedic Impairments: Roll the ball from a seated position. Also, a ramp can be used so that the student can simply place the ball at the top of the ramp and let it roll to the wall. Create barricades (using cardboard) to keep the ball within a controlled area.

Learning Disabilities: Place large folded cards with the numbers "1," "3," and "5" at each of the taped increments to provide clearer guides to distances.

Emotional Disabilities: Consider doing activity in pairs rather than larger groups if the commotion of the activity is overstimulating.

Closure

1. Was it easier to tell if you were at the 1-, 3-, or 5-meter mark? Why do you think this is so?
2. How is rolling the ball like a dolphin using echolocation? (The ball is simulating sound waves emitted from the dolphin). How is it different? (It is a ball and not a sound wave.)
3. SONAR (Sound Navigation and Ranging) uses the same principle of emitting sound waves (active sonar) and receiving the rebounded waves (passive sonar) as dolphins. What are some of the uses of SONAR? (Finding boats and submarines, fishing, locating sunken ships and treasure, and so on.)

Assessment

Have student teams create a diagram showing themselves as "dolphins," the path of the "sound waves" to the wall and back to them. Make sure students can relate the throwing of the ball to the emission of sound waves.

Note: This activity is loosely based on the resource guide *Bats Incredible! AIMS Activities 2–4*, available at *www.aimsedu.org/item/1125/Bats-Incredible-2-4/1.html*

Lesson 5

Meet Some Animal Helpers

To the Teacher

Many working animals provide assistance for humans. Service animals, such as dogs that help people who use wheelchairs or that guide blind people, perform tasks that the human is unable to do. Other animals, called therapy animals, help people who might need companionship, affection, or an audience for reading. These working animals allow people to function more independently and happily. In this lesson, students will read about four different animal helpers and share their ideas about the animal-human bond. Students will also attempt to teach a partner behavior without using words, reinforcing the idea that service-animal training takes tremendous time, patience, and commitment.

Objective

Students will be able to identify attributes of and activities associated with animals that help people. They will also be able to describe the process of learning new behaviors.

Time Needed

One 50-minute class period

Materials

The book *Animals at Work*, by Liz Palika and Katherine A. Miller, Animal Helper Question Sheet.

Teacher Preparation

Select four readings (or more depending on the number of groups in your class) about Service Animals from the *Animals at Work* book. Several short vignettes are given about guide dogs, service dogs, therapy cats, therapy horses and so on.

Procedure

1. Show students a photo of a guide dog. Ask, "What is this dog doing?" "Is it working?" "What kind of tasks does the dog do?" "How did it learn to do all of these things?"
2. Inform students that they are going to get to "meet" several service animals who help people with special needs.
3. Provide each group with a short reading about one of the service animals. Have students answer questions in groups.
4. Questions:
   (a) What kind of animal is your service animal?
   (b) How does the service animal help the human?
   (c) How did the service animal and the human get matched up?
   (d) What kind of training did the service animal need?
   (e) How does the human feel about the service animal?
5. Have student teams present on their animal/person partnerships.
6. Explain to students that they are going to role-play being trainers, service animals, and people who use service animals in order to learn about animal training.
7. Put students in groups of three. One student will be the trainer, one will be the service animal, and one will be the person who uses the service animal.
8. The trainer decides on a behavior he would like to teach the service animal (such as bringing an item back, stopping every time he sees a person, or barking when there is danger). The behavior can be complicated (multistep) but have just one command name such as "Bring!" or "Danger!"
9. The trainer must teach the service animal the behavior *without using words* other than the command word! They can use hand signals, head nods, and sounds, but not verbal instructions.
10. Once trained, the person who will use the service animal learns and gives the command word and must guess what the service animal's learned behavior is.

Accommodations for Students With Disabilities

Visual Impairments: Provide Braille or large-text readings; do not assume that the visually impaired student should be the person who needs a service animal; allow all students to select roles and change if desired.

Hearing Impairments: Use a visual command rather than a verbal command word.

Motor/Orthopedic Impairments: Behaviors can be stationary behaviors such as looking up and down, sticking the tongue out, and so on; do not assume that a student who uses a wheelchair should be the person needing assistance; allow all students to select roles and change if desired.

Learning Disabilities: Have students read text in a "round robin" manner and preview vocabulary words before reading.

Emotional Disabilities: If student has difficulty in a particular role, allow partners to continue training and allow student to reenter group as she or he feels comfortable.

Closure

1. How difficult or easy was it to teach and learn a behavior without words?
2. What were some successful ways of communicating?
3. Ask students to discuss training of their own pets. How do they communicate? Do they reinforce good behavior?

Assessment

Students can write letters to organizations that train service animals such as:

- The Delta Society: *www.deltasociety.org*
- Foundation for Pet-Provided Therapy: *www.loveonaleash.org*
- Guide Dogs of America: *www.guidedogsof America.org*
- Guide Horse Foundation: *www.guidehorse.org*
- Canine Companions for Independence: *www.caninecompanions.org*

Scoring Rubric

Criterion	Score	Points
Does the letter state the student's opinion on working animals?	____	5 pts
Does the letter provide at least three specific facts learned from this unit?	____	15 pts
Does the letter include information on the organization's work?	____	10 pts
Does the letter include a question for the organization?	____	5 pts
Does the letter contain proper grammar and structure?	____	5 pts

Total Pts. ____________/ 40

2

Lesson 6

How Do YOU Feel About Working Animals?

To the Teacher

Students have now been introduced to the lives of several types of working animals. For this activity, students will be asked to express an opinion about working animals by responding to questions about which animals they would and would not like to be. The purpose of this activity is to clarify each other's values about the issue of working animals, and to help students to understand that there are no wrong answers to these questions. However, students must "back up" their opinions with facts learned through the unit. This lesson also serves as a review of the material learned.

Objective

Students should be able to articulate their views on working animals and participate in respectful discourse on the topic.

Time Needed

One 50-minute class period

Materials

Four signs (each with a picture and name of a working animal, such as a guide dog, a draught horse, a performing elephant, a camel being ridden, and so on), tape

Teacher Preparation

Create four signs of working animals from photos on the internet. Tape each of the four signs up at the four corners of the room. (It's OK if the room is not perfectly angled, just so long as the signs are spaced all around the room.)

Procedure

1. Review the different types of working animals you have studied so far. Ask students, "Do you think YOU would like to be a working animal?" Accept all responses.
2. Explain to students that today, they are going to give their opinions on working animals. "What is an opinion?" "Is opinion the same as fact?" "Can there be more than one right opinion?"
3. Have students stand in the center of the room. Ask the question, "If you had to be a working animal, which of the following four animals would you rather be?" (Show the animal photos in the four corners of the room. Give students about a minute to think about it silently.
4. Instruct students to go to the corner of the room that is displaying their preferred working animal. Have students discuss with the other students there why they chose that one. (If a student is alone at a corner, join her or him and discuss.)
5. Allow groups to discuss for approximately five minutes. Then ask students to choose a spokesperson to report to the class on why the students chose their corner.
6. Allow spokespeople from each group to explain their group's reasons. Allow other students from the corner to contribute if they wish.
7. After each "corner" has reported, ask students, "Did you hear anything that surprised you?" Students may share that there were some ideas that they hadn't thought of or opinions that they didn't share but made sense.
8. Bring students back to the center of the room. Repeat the activity, but this time ask, "Which of these animals would you LEAST want to be?" Send students to corners and have them discuss and report.

Accommodations for Students With Disabilities

Visual Impairments: Make large-print signs on contrasting colors; use stuffed animals to give a tactile opportunity for students to connect with animal choices; clear pathways from center of the room to corners.

Hearing Impairments: If room is too large to ensure audibility, use the four corners of a large table to keep students in closer proximity. Use "pass the microphone" technique so only one student speaks at a time.

Motor/Orthopedic Impairments: Clear pathways between center of the room and corners; if movement is difficult, allow student to make her/his choice about corners and then place the photo of her or his animal on the desk, so that the student's desk becomes the corner.

Learning Disabilities: Allow students to take notecards and pencils if they would like to the corners so that they can record (writing or pictures) the information they are hearing; use words and pictures to designate the animal corners.

Emotional Disabilities: Assign student to the spokesperson role to help with focus; do the first activity as a tabletop activity (four corners of the table) to keep things more focused and calm; then do the second activity as a whole-room exercise.

Closure

1. How did it feel to hear the opinions of people who agreed with your choice of animal?
2. How did it feel to hear the opinions of people who did not agree with your choice of animal?
3. Can you name other examples of times when people have different opinions about something but there is no wrong answer?

Assessment

Students draw the working animals that they both did and didn't want to be and include a sentence explaining their opinion on each.

Lesson 7

Circus Animals

To the Teacher

Among the many groups of working animals are performing animals. Performing animals work in many different venues, including zoos, aquaria, circuses, and in the movies and television. Visitors to zoos and circuses are often unaware of the type of care and feeding required for these animals, many of which are large or exotic. In this lesson, students become familiar with the kinds of animals that perform in circuses, engage in a math activity on food requirements, and write a letter from the perspective of a circus animal about its daily living needs. The lesson can be extended by having students design a habitat for a circus animal.

Objective

Students will be able to identify the basic needs of animals including food, water, shelter, air, and companionship.

Time Needed

One class period (plus one extra class period if optional extension activity is done)

Materials

Book *Peter Spier's Circus*, by Peter Spier (Dragonfly Books 1995), Pros/Cons handout, "How Many Pounds of Food Do I Eat" handouts, pencils.

Procedure

1. Ask students, "Have you ever been to or seen pictures of a circus?" "What would you expect to see at a circus?" (jugglers, clowns, acrobats, animals). "Let's read a book and see if we can spot any working animals at a circus."
2. Read the book, *Peter Spier's Circus*, by Peter Spier. Ask, "What kinds of animals did we see in this book?" (horses, lions, monkeys, bears). Are they

working animals? What are their jobs in the circus? (to perform). Ask students, "Do you think that animals should perform in circuses?" (accept all answers and encourage students to give reasons for their answers).

3. "Imagine that you are in charge of taking care of the animals at a circus. What kinds of things would you need to know about animals?" (i.e., how much they eat and drink, how much water they need, space, friends, temperature, and so on).
4. Explain that today we are going to focus on how much circus animals eat, while next class, we'll learn more about how they are trained and cared for.
5. Provide students with "How Many Pounds of Food Do I Eat?" activity sheets. Explain that each of the four animals (elephant, tiger, monkey, and bear) require a certain amount of food each day to survive. The students are challenged to guess how many pounds of food each animal eats and then figure out how many of the representative food items that would equal, and circle them. Inform students that the animals don't actually eat *only* those foods; the foods shown on the sheets are just representing a certain number of pounds of food each day.
6. Students should work with teammates to discuss their guesses and calculations.
7. Review student responses and provide answers (Elephant: 150; 15. Tiger: 80; 16. Monkey: 6; 6. Bear: 40; 20) Ask, "Are you surprised by any of the answers? If so, which ones?"

Extension

Have students choose a circus animal and design an "ideal" home for it. What would it look like? How big would it be? How would the animal travel between circus destinations? Who would live with it? Use the National Zoo's "Design a Panda Habitat" at *http://nationalzoo.si.edu/education/conservationcentral/design/daph_broadband.htm* for ideas about what goes into keeping animals in a zoo, and how some of those ideas might, and might not, translate to circus animal keeping.

Assessment

Students will pretend they are an animal performer and will write a letter to their trainer/caretaker/director about its needs for each day. Student responses should include reference to food, water, veterinary care, other animals, exercise, fresh air, and kind treatment. The letter template is on page 128.

Scoring Rubric

Is the letter written from the perspective of an animal performer?	____ 5 pts
Does the letter include reference to at least five needs that the animal is requesting are met?	____ 15 pts
Does the letter contain proper grammar and structure?	____ 5 pts

Total Pts. ___/ 25

Name: ______________________________ Date: ______________

Animal Appetites Worksheet

How many pounds of food do you think an elephant needs each day to survive? Circle the bags of peanuts that you think an elephant needs for one day. Each bag represents 10 pounds of food.

How many pounds of food does an elephant actually eat each day? ___
Bags of peanuts ______

Name: ______________________________ Date: ____________

How much meat do you think a full grown tiger will eat in one day? Circle the amount of meat that you think the tiger will eat in one day. Each steak represents 5 pounds of meat.

How many pounds of food does a tiger actually eat each day? ______
Number of steaks _______

Name: ______________________________ Date: ______________

How much food do you think a monkey eats each day? Circle the amount of food you think the monkey eats each day. Each banana represents 1 pound of food.

How many pounds of food does a monkey actually eat each day? _____
Number of bananas? _____

Name: ______________________________ Date: ______________

How much food do you think a fully grown bear will eat each day? Circle the amount of food a bear will eat in each day. Each fish represents 2 pounds of food.

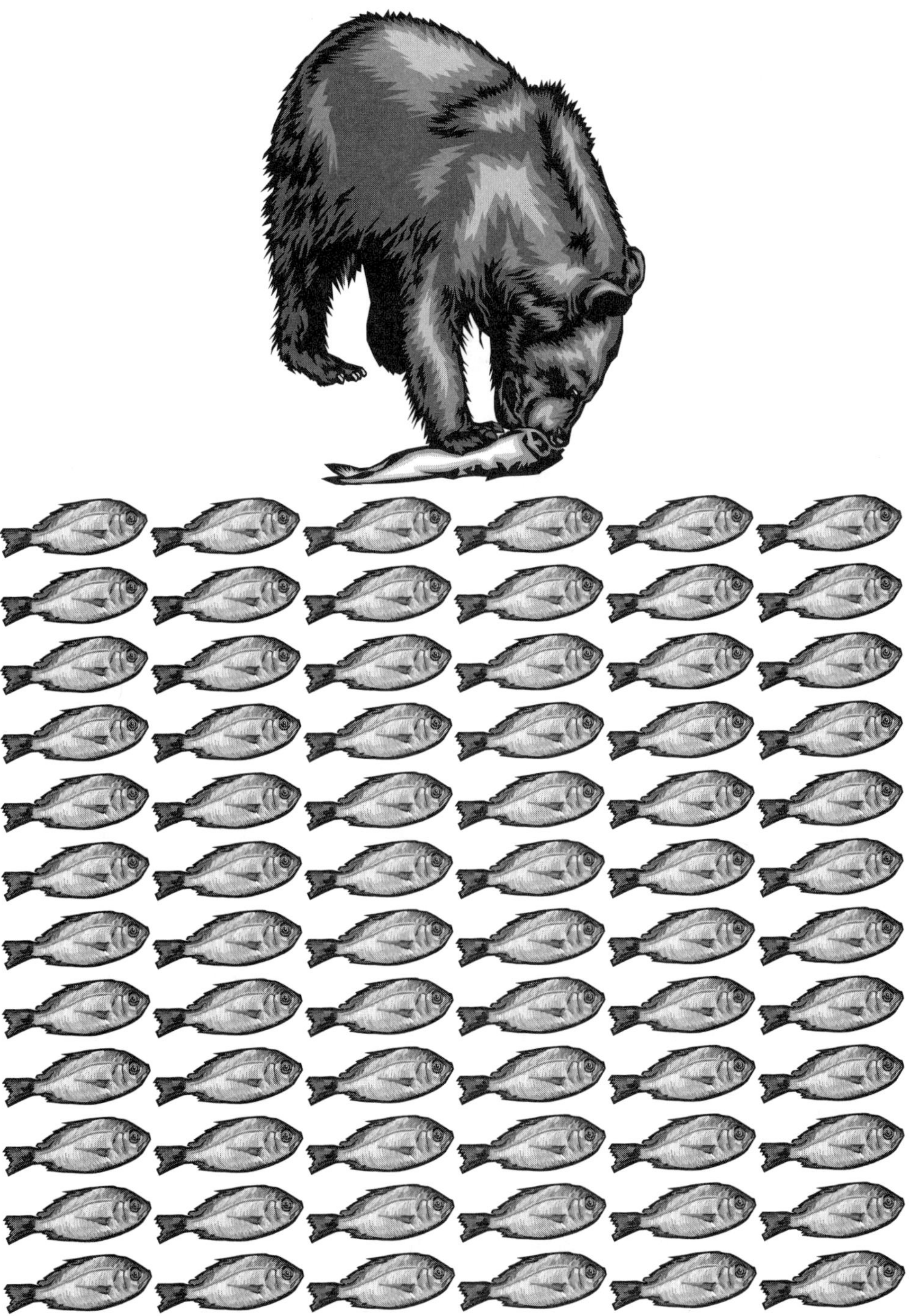

How many pounds of food does a bear actually eat each day? ______
Number of fish? _____

Pretend you are an animal performer. Write a letter to your trainer/caretaker/director explaining your needs for each day.

Date: ______________________________

Dear________________________________,

____

____

____

____

____

____

____

____

____

____

____

____

____

____

____

____

____

Sincerely,

____

2

Lesson 8

Under the Big Top

Animals in the Circus Debate

To the Teacher

Since its earliest times, the circus has been a popular source of entertainment and tradition. Circuses often include acrobats, clowns, tightrope walkers, trapeze artists, and performing animals. Many animal rights groups are opposed to circuses that use animals, as they believe the animals are abused and forced to perform against their will. Circus owners and trainers contend that the animals are treated humanely. While circus animals are regulated under the United States Department of Agriculture (USDA), who inspects the animals and their enclosures, many reports have surfaced of the years about mistreatment or abuse, causing some to question whether animals should be involved in circuses at all. During this activity, students will hear from and read about various stakeholders in the debate and will begin to form questions, seek out answers, and form opinions on this highly controversial issue.

Objective

Students will be able to interpret and analyze information from readings and construct arguments to advance their claims using evidence.

Time Needed

One 50-minute class period

Materials

Handout "Should Animals Perform in Circuses?" Expert question sheet, pencils

Teacher Preparation

This activity uses a Jigsaw strategy to help students read about the various players involved in the circus animal controversy and to begin to let them take on the various roles. If students normally belong to a set group or table, they can remain at those tables; they will be sent to "expert groups" with other students during the activity.

Procedure

1. Explain to students that they are going to read information about the different sides of the circus animal question and then discuss together.
2. Give each student one "Should Animals Perform in Circuses?" handout, which includes three articles.
3. Assign each student at a table to one of the stakeholders (circus owners, animal rights organization members, USDA inspectors, town council). As a class, read the entire page aloud.
4. Send students to "expert groups" so that all of the circus owners are at one table, all of the animal rights organization members are at another table, and so on. Instruct expert teams to reread their section to each other and answer the "expert team" questions. Town council members should be familiar with all of the readings.
5. Send students back to their original tables to share the answers from their expert teams.

Accommodations for Students With Disabilities

Visual Impairments: Prepare Braille or large-print handouts; describe the pictures in the book as you read.

Hearing Impairments: Use "round robin" technique for reading so that it is obvious who is reading and who reads next. Provide printed directions of Jigsaw process (i.e., home team → expert team → home team).

Motor/Orthopedic Impairments: Clear pathways between student's home team and expert team; allow student to record answers to questions if writing is not possible.

Learning Disabilities: Provide students with graphic organizers to sort facts; allow students to use highlighter pens to highlight important facts.

Emotional Disabilities: Some students may have difficulty being assigned to expert groups representing opinions with which they don't agree. Remind students that they are only focusing on that opinion in order to get information—they will be given a chance to voice their own opinions at another time.

Closure

1. What are some of main arguments FOR having animals perform in circuses? What are some the main arguments AGAINST having animals perform in circuses?
2. Do you need any more information to decide how you feel about this issue?
3. Do people ever disagree on important decisions? What are some ways people can work together to make good decisions?

Assessment

Have students write or verbally describe at least two arguments from an *opposing* viewpoint to their stakeholder. In other words, if the student has role-played a circus owner, they should provide arguments against circus animal performers.

Expert Team Questions

1. What is the name of your person or group?

2. Are they FOR or AGAINST having animals perform in circuses?

3. Give at least THREE reasons they feel the way they do.

4. What would you say to argue AGAINST them? List as many ideas as you can!

Should Animals Perform in Circuses?

Article in Support of Animals in Circuses: Circus Owners

Many circuses support the use of animals as performers. Circus owners explain that their businesses will only succeed if their animals are happy and healthy. They say that it would be bad for their business if they abused their animals.

Circuses claim that they provide food, water, veterinary visits, and clean and stimulating environments for their animals. They also explain that they make regular stops during long train trips so that the animals can exercise and socialize. In fact, many circuses state that circus animals only spend a small part of their day performing and that most of the day is spent eating, sleeping, and socializing with other animals.

Circuses explain that they train their animal performers using positive reinforcement. This means that the animals get rewards, such as food or praise, for doing tricks well. Animal trainers have the animals perform tricks many times so that they can learn them quickly and receive rewards. Circuses also claim that most of the animal tricks are part of the animal's natural behaviors that they display during play time.

Circuses point to the fact that many of their animals live longer than they would have in the wild, thanks to the care and safety they provide. Circuses also explain that they are regulated by the United States Department of Agriculture's Animal Welfare Act, which means that they get inspected regularly. Circuses can get fined or even forced to close if they do not pass their inspections.

Circuses believe that it would be wrong to ban animals from performing because animal acts have a long tradition in circuses and families have been coming to see the acts for many years. They also feel it would be wrong to ban animal performances because the trainers and animals care about each other. Many circus owners support the laws that regulate circus animal care and believe that circuses that neglect or abuse animals should be punished. But circus owners don't want all circuses to be banned just because some circuses break the rules.

Resources

Carson & Barnes Caring for Circus Animals: *www.carsonbarnescircus.com/caring-circus-animals*

Cole Bros. History: *www.gotothecircus.com/index.php/cole-bros-history-sp-1020991886*

Ringling Brothers Animal Care: *www.ringling.com/ContentPage.aspx?id=45760§ion=45696*

Article Against Animals in Circuses: Animal Rights Organization Members

Many animal rights organizations oppose using animals as performers in circuses. Most of the objections stem from their belief that circus animals, such as elephants, bears, lions, and tigers, do not do their tricks voluntarily; they do them because they are afraid of being punished.

One organization known as PETA (People for the Ethical Treatment of Animals) alleges that many circuses use whips, electric prods, bullhooks, and tight collars to force animals to do tricks. They also suggest that some circuses start training baby elephants when they are too young and scared to fight back.

Animal rights organizations also claim that circus animals are confined in small cages and are often forced to travel long distances in train boxcars without fresh air, water, or food. They also claim that the constant traveling means that animals are chained up for many hours at a time, without an opportunity to exercise or move freely. The animal rights organizations point to incidences where circus animals have died in train boxcars from heatstroke in order to convince the public that circus animals suffer from unsafe travel conditions.

Animal rights organizations also feel that keeping wild animals in circuses is dangerous for the trainers and the public. Animal rights organizations point to examples where circus animals have hurt spectators. Sometimes, unfortunately, those animals have to be killed.

Organizations like PETA obtain photos of animals being treated cruelly in circuses and then show the photos to the public to try to get them to stop going to circuses. They encourage people to ban circuses that use animal performers. They also encourage the public to support animal-free circuses and other forms of animal-friendly entertainment. Many countries, including the United Kingdom, India, Greece, and Bolivia, have banned the use of wild animals in circuses.

Resources

Animal Defenders International (ADI). "Greece Bans Animal Circuses" *www.ad-international.org/animals_in_entertainment/go.php?id=2528*

People for the Ethical Treatment of Animals (PETA): *www.peta.org/issues/animals-in-entertainment/circuses.aspx*

Ban on wild animals in circuses in the United Kingdom: *www.express.co.uk/news/nature/438753/AT-LAST-Ban-on-ALL-wild-animals-in-circuses-is-passed*

"Neutral" Circus Animal Reading: USDA Inspectors

Circuses, zoos, and marine mammal parks are regulated under the Animal Welfare Act. The Animal Welfare Act was signed into law in 1966 with the goal of protecting animals by developing minimum standards for their care and handling. The Act is enforced by the U.S. Department of Agriculture (USDA). The USDA's Animal and Plant Health Inspection Service (APHIS) performs annual surprise inspections on circuses to ensure that the circuses are following the rules of the Animal Welfare Act. If a circus is violates the rules, it can be fined and more frequent inspections can occur.

Inspectors check many things during inspections including how animals are trained and handled, whether their cages and traveling enclosures are large enough and provide enough air and light, and whether animals receive enough food, water, and rest between performances.

If a circus violates any of the regulations, it is given time to fix the problem. Then, an inspector returns to the circus to check whether the problem was fixed. If the circus repeatedly violates a regulation, it is fined. The USDA can even take circuses to court to try to have them closed if the violations are serious enough.

Some animal rights organizations claim that these types of inspections cannot properly monitor the treatment of animals. They are particularly concerned about traveling circuses since it is harder for government inspectors to follow-up on violations when a circus is constantly moving around. Circus owners, on the other hand, say that it is in their best interest to follow the regulations because they want their animals to be happy and healthy and able to perform.

Resources

Newsday Article on Circus Animal Regulation: *www.newsday.com/long-island/nassau/how-the-government-regulates-treatment-of-circus-animals-1.1210151*

U.S. Dept. of Agriculture. National Agriculture Library: *http://awic.nal.usda.gov/zoo-circus-and-marine-animals*

U.S. Dept. of Agriculture. Animal Welfare Act: *www.aphis.usda.gov/animal_welfare/downloads/Animal%20Care%20Blue%20Book%20-%202013%20-%20FINAL.pdf*

Lesson 9

Where Do You Stand?

To the Teacher

During this activity, students will "take a stand" on the issue of whether animals should perform in circuses. During the first part of the activity, students will use a "spectrum" line to show how they feel about the issue. They will engage in discussion with classmates, who will each express their opinions on the issue. Then, students will continue research on the topic and will present their arguments through various modes of expression.

Objectives

Students with synthesize research in order to articulate their own perspective on the issue of whether animals should perform in circuses.

Time Needed

Three 50-minute class periods

Materials

Masking tape, two pieces of construction paper, computer with internet access, Skype or other videoconferencing platform

Teacher Preparation

Place a long line of masking tape along the floor from one side of the room to the other. On one side of the room, place a piece of construction paper that says (or shows) "Performing Animals Should Stay in Circuses!" and another on the opposite side that says (or shows) "Performing Animals Should Be Banned From Circuses!"

Procedure

Day 1

1. Ask students to get into groups according to the last class's activity (circus owners, USDA inspectors, animal rights organization members, town council).
2. Have each team report on their viewpoints as a reminder to the class.
3. Explain to students that what they have been doing is called, "role-playing"; that is, they have been taking on the opinions of others as their own.
4. Explain to students that today, they are going to have a chance to share their own opinions on the issue. Remind students that there are no right or wrong opinions, but students should be able to explain their opinions by giving reasons.
5. Show students the "spectrum" line across the room.
6. Instruct students to determine where they belong on the line when you ask the question, "Should animals perform in circuses?" Students can go to either end of the spectrum, or anywhere in between (in the event students are not sure or have mixed feelings).
7. When students are in positions, ask them to share the main reasons they are where they are.
8. After discussion, ask students if they wish to move anywhere on the line. If so, ask what persuaded them to move.

Day 2

1. Inform students that they will have the opportunity to choose a project that will allow them to share their opinions about performing animals in circuses. The options include:

   (a) Development of a cartoon strip that includes dialogue between two stakeholders (i.e. animal and trainer, animal rights organization member and town council member, and so on);

   (b) Development of a puppet show depicting a scene between issue stakeholders;

(c) Development of a meeting at City Hall where the City Council will vote on the issue.

2. Explain to students that you will be looking for the arguments that support the opinions of each of the characters. What "evidence" do they have for what they say? Can resolutions or compromises be reached between the characters?
3. Allow students to work on their projects alone or in groups.
4. Have students present their projects.

Accommodations for Students With Disabilities

Visual Impairments: Give verbal directions such as "right" and "left" during spectrum line activity. Allow student to create a podcast if desired.

Hearing Impairments: Use "pass the microphone" technique in team activity; make sure students face hearing-impaired student during spectrum line activity if lip-reading.

Motor/Orthopedic Impairments: Clear pathways between sides of the line. If students have difficulty moving, allow them to use a flashlight to shine on their chosen spot.

Learning Disabilities: Use pictorial and text cues on the pictures (animals/no animals).

Emotional Disabilities: If the spectrum line proves to be overstimulating as a whole-class activity, allow students to do this as a tabletop activity with tape across the table and students placing bingo chips on the position of their preference.

Closure

1. What are some of main arguments FOR having animals in circuses? What are some the main arguments AGAINST having animals in circuses?
2. Are we able to reach any agreement about this issue? Is it possible to compromise?

Assessment

Evaluate student projects using the following criteria:

1. Did student state the opinions of the stakeholders?
2. Did student back up the opinion with supporting evidence?
3. Did student demonstrate understanding/awareness of counterarguments?
4. Did student articulate/express a resolution between the stakeholders?

Scoring Rubric

	1 pt.	2 pts.	3 pts.
Use of Evidence	Student uses opinion without evidence or inaccurate evidence to back claims	Student uses tenuous or incomplete evidence to back claim	Student demonstrates complete and accurate use of evidence to back claims
Opinions of Stakeholders	Student does not recognize or distinguish between various stakeholders	Student identifies the opinion of one stakeholder	Student accurately states the opinion of all stakeholders
Science Content Understanding	Student demonstrates minimal understanding of science content	Student demonstrates a moderate degree of understanding of science content	Student demonstrates strong understanding of science content and consistently applies it to argument
Argumentation	Student is unable to formulate counterarguments or resolutions	Student formulates simplistic counterarguments or resolutions	Student demonstrates skilled abilities in formulating counterarguments or resolutions

Lesson 10

Culminating Project

A Whole-Class e-Book on Working Animals

To the Teacher

As a culminating activity to the unit, students will develop a whole-class e-book to which each student will contribute a page about a working animal.

Objective

Students will be able to identify, describe, and communicate information about a working animal and present it in an electronic format.

Time Needed

One 50-minute class period

Materials

Computers with access to any software capable of allowing students to draw or design a page with text. (Sound capabilities are needed if you would like the students to narrate.) Options include PowerPoint and Photo Story.

Teacher Preparation

Determine software to be used and have computer(s) accessible for class.

Procedure

1. Ask students their thoughts about the "working animals" unit. Did they enjoy it? Did they learn anything new? Did their opinions on the subject change at all?
2. Explain that today they are going to share some of their knowledge with the community through the creation of an online e-book about working animals.

3. In the whole group, have students identify a fact that they would like to share in the book. Record on the board which student wants to share which fact. Try to encourage students to discuss many of the different types of working animals they learned about.
4. Once each student has identified their "page" of the book, students go to computers/laptops/iPads and create a page that includes a picture and a sentence for the book.
5. Once all students have completed their page, compile their pages into one file and add narration, if desired.
6. Upload the e-book on the school website (and on parent-accessible website) for distribution!

Accommodations for Students With Disabilities

Visual Impairments: Provide computer with auditory command capabilities.

Hearing Impairments: Include written text if book is narrated.

Motor-Orthopedic Impairments: If adequate assistive technology is not available for student to draw online, allow student to create his or her page on paper and scan it for electronic presentation in the book.

Learning Disabilities: Provide students with a written word bank for cuing and spelling.

Emotional Disabilities: If selection of page topic proves to be a challenge, randomly assign students to pages.

Assessment

Students are assessed on the quality and comprehensiveness of their e-book page.

Scoring Rubric

Did the e-book page clearly identify a working animal?	____ 5 pts.
Did the e-book page include at least three facts about the animal?	____ 15 pts.
Did the e-book page include an illustration or photo?	____ 10 pts.
Did the student use appropriate technology for the e-book page?	____ 5 pts.
Was the e-book page creative?	____ 5 pts.

Total Score: ___/40 pts.

Unit 3

A Need for Speed?

Should speed limits be lowered to reduce traffic fatalities?

Unit Overview

During this unit, students will investigate physical science concepts related to forces and motion. They will then synthesize and apply their learning to grapple with the question of whether speed limits should be lowered to reduce traffic accidents. Through engagement in several hands-on activities, as well as the development of research-based arguments, students relate physical concepts to their everyday lives and to the functioning of society.

Key Science Concepts

Forces, Friction, Motion, Momentum, Mass, Gravity, Velocity

Ethical Issues

Personal Freedom, Government Regulation, Individual vs. Societal Interests

Science Skills

Predicting, Observing, Measuring, Analyzing Data, Understanding Cause and Effect, Communicating Results, Forming Arguments From Evidence.

Grade Levels

Upper Elementary and Middle School

Time Needed

The unit is comprised of four lessons over five class periods. Time can be adjusted depending on student content background.

Lesson Sequence

Lesson 1. Slippery When Wet (Momentum and Friction on the Slip 'n Slide)

Lesson 2. Data Driven (Analyzing and Interpreting Slip 'n Slide Data)

Lesson 3. Speed Kills? (Research on Speed and Automobile Safety: 2 periods)

Lesson 4. Town Hall Meeting: Should We Reduce Speed Limits?

Background on the Issue

Speed limits are part of everyday life in societies that rely on motor vehicles for transportation. Yet questions about whether speed limits really do save lives, and if so, what speeds are "ideal" remain controversial. Speed limits are generally imposed for safety reasons, but also have been used to reduce fuel consumption and reduce air and noise pollution. U.S. federal highway speeds have changed over the years, in response to research on road safety as well as fuel prices. Those who favor reduced speed limits generally cite research suggesting fewer road fatalities, improved fuel efficiency, and cleaner environments. Those who oppose speed limit reductions generally cite research suggesting that speed limits do not substantially improve safety or reduce air pollution, and that they are irrelevant since many people don't obey them anyway. Furthermore, opponents suggest that speed limits make commerce more difficult as they slow the movement of goods for businesses, yet create a large source of revenue for city, state, and local government through ticketing. In this unit, students will explore how velocity and mass affect momentum and will use their knowledge to analyze and interpret readings on speed limits and road safety. Students will then debate the proposed reduction of speed limits in a town hall meeting using evidence informed by their research.

Connecting to *NGSS*

3-PS2-1. Plan and conduct an investigation to provide evidence of the effects of balanced and unbalanced forces on the motion of an object.

3-PS2-2. Make observations and/or measurements of an object's motion to provide evidence that a pattern can be used to predict future motion.

MS-PS2-2. Plan an investigation to provide evidence that the change in an object's motion depends on the sum of the forces on the object and the mass of the object.

Accommodations for Students With Disabilities

Visual Impairments: Allow student to tactilely examine the Slip 'n Slide in advance of the lesson; have student count the number of paces back from the Slip 'n Slide so that she or he knows when to slide; allow student to feel the difference between one pound and one kilogram.

Hearing Impairments: Outdoor activities can be tricky as wind can increase "noise" for students who use hearing aids or cochlear implants; if possible, use an FM system or interpreter, or try to give directions either inside the school or in a wind-protected area if it happens to be windy; provide visual cues for vocabulary words such as *velocity*, *momentum*, and *mass*.

Learning Disabilities: Show students several examples of pound to kilogram conversions; separate the velocity and momentum portions of the data sheet into two pages to allow for greater white space on the page; use a T-chart to keep track of arguments for and against reduced speed limits.

Motor-Orthopedic Impairments: Accommodations will depend on the student's particular impairment and comfort level; speak with the student to discuss options for the Slip 'n Slide portion of the unit; the student could be seated on the Slip 'n Slide and have partners gently pull the student so that he or she can feel the difference in movement on a dry, wet, or soapy surface; if the student uses a wheelchair, the same calculations for velocity and momentum can be done using the wheelchair motion rather than the Slip 'n Slide; a whole-class alternative activity could consist of all students testing their speed using a spare wheelchair so that all students are using the same apparatus.

Emotional Disabilities: Assign numbers to students in advance of approaching the Slip 'n Slide to minimize arguments over the order of sliding; anticipate an outdoor "time out" spot that a student could move to if she or he requires a break from the excitement; if a student has difficulty listening or becomes disruptive during the debate, consider providing her or him with a Town Hall Meeting Participation Sheet and allow student to serve as an assistant to you; this may help student to focus on the conversation and provide a sense of autonomy.

Resources for Teachers

- The Physics Classroom on Momentum (*www.physicsclassroom.com/Class/momentum/u4l1a.cfm*)
- Insurance Institute for Highway Safety's Speed Limits Q&A Page: (*www.iihs.org/research/qanda/speed_limits.aspx*)

Note: This unit is based on a project submitted by Thomas Dolan.

An alternative to the "Slip 'n Slide" lesson that uses a skateboard can be found at:

Dolan, T. J., and D. L. Zeidler. 2009. Speed kills! (Or does it?). *Science and Children* 47 (3): 20–23.

3

Lesson 1

Slippery When Wet

To the Teacher

In this lesson, students will participate in a Slip 'n Slide activity to examine concepts of velocity, mass, momentum, and friction. They will record their time sliding on water with and without dish soap, calculate their momentum based on their weight, and examine the effects of a reduction of friction on their velocity.

Objective

Students will demonstrate their understanding of the impact of friction on velocity.

Time Needed

One class period

Materials

Slip 'n Slide, hose, stopwatch, bottle of dish soap, pencil, list (with the names of all students and 6 columns for each student), bathing suits, towels, goggles (1 pair per student).

Note: You may wish to have an extra adult on hand in order to assist with the data collection.

Procedure

1. In advance of the class, set up the Slip 'n Slide outdoors.
2. Bring students outside to the Slip 'n Slide and ask for a volunteer to slide down the track without water turned on. (Students will either refuse because they know they won't slide, or will sit on the Slip 'n Slide without moving.)

3. Probe students about friction. "Why can't we slide well on the dry Slip 'n Slide?" "What is friction?" "How can we reduce the friction on the track?"
4. Have students line up and spray them with the hose. Students will then line up and run a distance of 5 meters before proceeding headfirst down the Slip 'n Slide track. The Slip 'n Slide track is approximately 6.1 meters in length.
5. Have a parent or another adult time students on the track with a stopwatch and record student times on the data sheet.
6. Allow each student to slide three times with the water running.
7. After each child has gone three times, apply dish soap to the Slip 'n Slide. (This is used to further demonstrate the lessening of friction.) Have students wear goggles and allow them to take another three turns down the slide with the soap added. Record times.
8. After all students have gone a total of 6 times, students will rinse off and proceed to change back into their school clothes.

Slip 'n Slide Time Sheet

Enter the number of seconds for each student's trials
(trials 1–3 with plain water, trials 4–6 with soapy water)

Student	Trial 1	Trial 2	Trial 3	Trial 4	Trial 5	Trial 6

Lesson 2

Data Driven

Topic: Momentum
Go to: *www.scilinks.org*
Code: SSI005

To the Teacher

During this lesson, students analyze and interpret their results from the Slip 'n Slide activity to draw conclusions about the relationship between velocity and mass to momentum, as well as the impact of friction on velocity.

Objective

Students will demonstrate that both mass and velocity are related to momentum, and that friction reduces velocity.

Time Needed

One class period

Materials

Slip 'n Slide handout, timesheet (recorded from the previous day's activity for each student), pencil, paper, calculator.

Procedure

1. Begin by reviewing the events of the previous class. Can we slide on the Slip 'n Slide without water? (no, because too much friction) What is friction? (a force that slows things down) How did the water help us slide? (reduced friction). What effect do you think the soap had? (made it more slippery and reduced friction) Is it easier to stop with or without soap? (without)
2. Explain to students that in order to be certain about our results, we need to calculate our velocity, which is a function of time and distance. Have students examine their data sheets.

3. Students complete their data sheets by entering their mass (converted to kg as per the directions on the sheet) and then calculate their velocities both with and without soap. Students also calculate their momentum, which is a function of velocity and mass.

4. Have students compare and contrast the data in order to determine the effects of mass and velocity on momentum with regard to variances in friction due to the water and soap (soap decreases friction which then increases velocity; increased velocity increases momentum). Also note that velocity is inversely proportional to time for a given distance; in other words, it takes less time to get to a destination when traveling faster.

Closure

Pool the class data so students can compare their results with the class to look at the effect of mass on momentum (increased mass increases momentum). Students should observe that increasing both mass and velocity increase the momentum of an object. Note that an object with greater momentum takes longer to stop.

Assessment

Ask students to determine what factors would make a vehicle more or less likely to come to a quick stop, based on our results. (a fast, heavy vehicle will have more momentum than a slow, lighter vehicle, and will therefore stop more slowly; wet or oily roads reduce friction so velocity and momentum are increased)

Name: ______________________________ Date: ____________

Slip 'n Slide

Mass Conversion (Round answer to the nearest hundredths)

Your Weight (lbs.) × 0.454 = Your Mass (kg)

____________ × 0.454 = ____________

Time Trials (Round time to the nearest tenth of a second)

Without Soap Trials

Trial	Time
1	
2	
3	

With Soap Trials

Trial	Time
1	
2	
3	

Velocity Calculation (Round answers to the nearest hundredths)

Trial	Distance (m) ÷	Time (sec) =	Velocity (m/sec)
1			
2			
3			
4			
5			
6			

Momentum Calculation (Round answers to the nearest hundredths)

Trial	Mass (kg) ×	Velocity (m/sec) =	Momentum (kg m/sec)
1			
2			
3			
4			
5			
6			

Lesson 3
Speed Kills?

To the Teacher

In this lesson, students will research various sources of data to grapple with the question of whether speed limits in the fictional city of Driversburg should be reduced in order to cut down on the number of vehicle fatalities.

Topic: Speed Limits
Go to: *www.scilinks.org*
Code: SSI006

Objective

Students will analyze a series of articles to become familiar with statistics and information regarding speed and the effect it has on vehicular accidents.

Time Needed

Two class periods

Materials

Article packet (containing 4 articles), Article Evaluation Sheets, highlighters, pencils

Procedure

Day 1

1. After reviewing some of the major concepts from the prior lessons (momentum, velocity, mass), ask students whether they think that there should be speed limits on motor vehicles? (accept all answers) Ask, "If you think there should be limits, how can we decide what those limits should be? Should cars just go as slowly as they can?" (accept all answers)
2. Inform students that they are going to investigate research that has been done on vehicle safety to determine whether speed limits should be reduced.

3. In groups of four, have students read the articles and highlight important information within each article. After all students have finished reading the articles, each group will individually discuss the effects of lowering the speed limit and complete the Article Evaluation Sheets.

Day 2

1. Ask students to imagine that they are citizens of the city of Driversburg. After a series of recent fatal vehicular accidents, a law was proposed to reduce the speed limits on city streets by half. They have been charged with debating the law and ultimately, voting on its outcome.
2. Place students in new groups (different than their research groups) that represent various stakeholders on the issue of speed limit reduction. The new groups are: truck drivers, parents, business leaders, and police officers.
3. Challenge students to use the information in the articles and the knowledge gained in the hands-on experiments to formulate arguments for their particular stakeholder group. Allow students to access additional resources, if necessary, to enhance their arguments.

Closure

Inform students that they will be debating the issue in the next class period. "What are our ground rules for our debates?" (raising hands to speak, listening respectfully, no personal attacks or criticisms, using evidence to justify claims, and so on)

Assessment

Review student article summaries for accuracy and completeness.

Articles for Packets

Article 1: Witzenburg, G. (2003, January 23). Seven major myths of speed and speed enforcement. *Consumer Guide. www.motorists.org/ma/myths.html*

Article 2: U.S. Department of Transportation Research, Development, and Technology. 1992. Fact Sheet: Effects of raising and lowering speed limits. *www.fhwa.dot.gov/publications/research/safety/humanfac/rd97002.cfm*

Article 3: Nagourney, E. 2009. Safety: As speed limits rise, so do death tolls. *The New York Times*. July 20. *www.nytimes.com/2009/07/21/health/research/21safe.html?scp=3&sq=highway&st=cse&_r=0*

Article 4: Reinberg, S. 2009. Deaths, injuries increase with higher speed limits. *U.S. News and World Reports* (online version). July 16. *http://health.usnews.com/health-news/managing-your-healthcare/articles/2009/07/16/deaths-injuries-increase-with-higher-speed-limits*

Name: ______________________________ Date: ____________

Article 1

"Seven Major Myths of Speed and Speed Enforcement"

Summarize the main argument of the article:

____

____

____

____

Identify key points used in this article:

1. ______________________________

____

____

2. ______________________________

____

____

3. ______________________________

____

____

4. ______________________________

____

____

5. ______________________________

____

____

Name: ______________________________ Date: ____________

Article 2

"Fact Sheet: Effects of Raising and Lowering Speed Limits"

Summarize the main argument of the article:

Identify key points used in this article:

1.

2.

3.

4.

5.

Name: ______________________________ Date: ______________

Article 3

"Safety: As Speed Limits Rise, So Do Death Tolls"

Summarize the main argument of the article:

____

____

____

____

Identify key points used in this article:

1. ______________________________

____

____

2. ______________________________

____

____

3. ______________________________

____

____

4. ______________________________

____

____

5. ______________________________

____

____

Name: ______________________________ Date: ______________

Article 4

"Deaths, Injuries Increase With Higher Speed Limits"

Summarize the main argument of the article:

____

____

____

____

Identify key points used in this article:

1. ______________________________

____

____

2. ______________________________

____

____

3. ______________________________

____

____

4. ______________________________

____

____

5. ______________________________

____

____

Lesson 4

Town Hall Meeting

Should We Reduce Speed Limits?

To the Teacher

In this lesson, students will use the evidence they have collected from their hands-on investigations and their article research to debate whether speed limits should be reduced in the fictitious city of Driversburg.

Objectives

Students will use evidence-based argumentation while engaging in respectful discourse on the controversial issue of whether speed limits should be lowered to reduce traffic accidents.

Time Needed

One class period

Materials

Completed article evaluation sheets and Slip 'n Slide data sheets from prior lessons, pencils.

Procedure

1. Organize students to participate in a town hall meeting to determine if the speed limit should be lowered to help reduce traffic fatalities in Driversburg. Truck drivers and business leaders will argue against the reduction in speed limit, while parents and police officers will argue in favor of the speed limit reduction. The final decision regarding the speed limit will be determined by a secret ballot at the conclusion of the debate.

2. Begin a semistructured debate whereby all groups make an opening statement comprised of one or two claims using evidence to persuade others of their viewpoint. After all four groups have made their statements, allow students to ask questions of other groups' evidence, refute their claims, or bolster their own claims with additional evidence. Students must raise their hands to participate.

3. At the end of the meeting, review your impressions with the students and clear up any misconceptions that may have surfaced during the discussion. Have students now assume the role of the governing body and vote using a secret ballot to determine whether the speed limit should be reduced in Driversburg. Students should vote what they believe, not their stakeholder positions (although these might be the same). Announce the results to the class.

Closure

Discuss the outcome with the students. "Were you surprised by the group's decision?" "Do you still have questions about the relationship between speed and vehicle safety?" "Could we have negotiated a different outcome?" "How does it feel to use evidence to inform decision making?" "Did you change your mind about this issue during the debate?

Assessment

Students write a letter to their town, county, or state department of transportation describing their research results and expressing their opinions on speed limits. Letters can contain additional research on local speed limits and speed-related accident statistics. Letters are assessed on accuracy of data, clarity of expression, persuasiveness of arguments, and format/grammar. Students can also be assessed on participation in all phases of Town Hall Meeting activities (See Town Hall Meeting Participation Sheet).

Name: ______________________________ Date: ______________

Town Hall Meeting Participation Sheet

(Place a check "√" in each box as task is completed.)

Student Names	Completed Article Research Sheets	Participated in Stakeholder Group Discussions	Made Evidence-Based Contribution to Town Hall Meeting	Developed a Thoughtful Letter to Department of Transportation	Total (4 "√" Max)
Truck Drivers					
Parents					
Business Leaders					

Name: ______________________________ Date: ______________

Student Names	Completed Article Research Sheets	Participated in Stakeholder Group Discussions	Made Evidence-Based Contribution to Town Hall Meeting	Developed a Thoughtful Letter to Department of Transportation	Total (4 "√" Max)
Police Officers					

Letter Scoring Rubric

Did the letter state the student's opinion on the issue of speed limits?	_____ 5 pts.
Did the letter provide at least five specific facts and/or statistics learned from this unit?	_____ 15 pts.
Was the letter accurate?	_____ 5 pts.
Was the letter clear?	_____ 5 pts.
Was the letter persuasive?	_____ 5 pts.
Does the letter contain proper grammar and structure?	_____ 5 pts.

Total Pts. _____/ 40

Unit 4

Space Case

Do humans have the right to colonize and use resources on extraterrestrial planets?

Unit Overview

During this unit, students will investigate the hierarchical relationships between planets and other astronomical bodies relative to the universe, including distance, size and composition. By exploring the controversial issue of whether humans have the right to utilize and colonize extraterrestrial environments, students will consider the requirements of life, the manner in which different resources are allocated and utilized, the dynamics of international cooperation on Earth, and the ethical considerations of impinging on the rights and resources of other potential life forms in space.

Topic: Space Ethics
Go to: *www.scilinks.org*
Code: SSI007

Key Science Concepts

Planetary Properties (including Geological Features, Orbitals, Atmosphere), Space Exploration, Requirements for Habitability, Resource Conservation, Sustainability

Ethical Issues

Allocation of Resources, Public vs. Private Property Rights, Colonialism, Conservation vs. Commercialism

Science Skills

Inferring, Analyzing Data, Hypothesizing, Developing Models, Determining Cause and Effect, Using Evidence to Form Arguments, Communicating Ideas

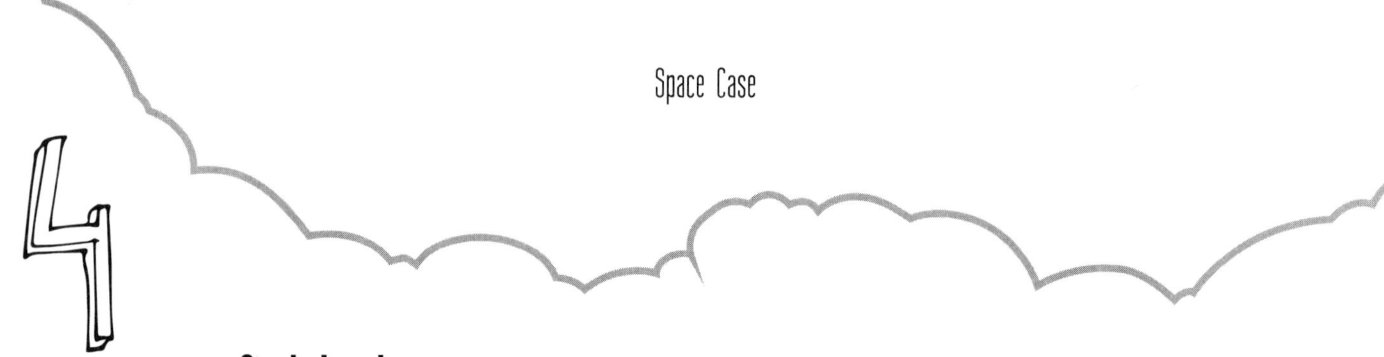

4

Grade Levels

This unit is appropriate for middle or high school physical science, astronomy, or environmental science classes.

Time Needed

This unit takes approximately two to three weeks, depending on the number of class meetings and the level of research.

Lesson Sequence

Lesson 1. Who Owns Outer Space? (Video and document research using Jigsaw)

Lesson 2. Space Players (Role-play a meeting between the United Nations, national agencies, settlement companies, and special interest groups)

Lesson 3. Planetary Particulars (Developing a detailed model and poster of their planet and completing an application for settlement)

Lesson 4. United Nations Vote (Presentation of planet proposals and United Nations vote on approval or denial of planetary settlement application)

Background on the Issue

> The goal isn't just scientific exploration … it's also about extending the range of human habitat out from Earth into the solar system as we go forward in time … In the long run a single-planet species will not survive. We have ample evidence of that [species have] been wiped out in mass extinctions on an average of every 30 million years.
>
> —Michael D. Griffin, NASA Administrator, to Washington Post, September 25, 2005

> Intelligent civilizations are out there and there could be thousands of them, according to an Edinburgh scientist. The current research estimates that there are at least 361 intelligent civilizations in our Galaxy and possibly as many as 38,000.
>
> —*International Journal of Astrobiology* (February 5, 2009)

Space Case

In 1959, the United Nations created the Committee on the Peaceful Uses of Outer Space (COPUOS). Since then, COPUOS has negotiated five major international treaties in order to identify the rights, responsibilities, and prohibitions of Earth's nations in regard to the use of space resources and environments. Although these treaties cover many issues including the prohibition of any one nation to appropriate outer space, the freedom of exploration, liability associated with space exploration, arms control, and the like, many questions still remain regarding: (a) the rights of humans to use extraterrestrial and extra-solar resources; (b) the question of whether settlements on extraterrestrial bodies constitute colonization; (c) whether extraterrestrial bodies can be established as private properties; (d) whether extraterrestrial settlements and resource usage are worthy economic investments; and (e) the processes by which humans determine who has a right to settle. These questions are highly relevant to students' lives as they address questions of space law, resource use, and human and property rights.

To address these issues, students will perform extensive research in order to create their own planet and settlement. This will require determination of: (1) Orbital distances, satellites, solar type, and planetary primary element composition; (2) Habitability, renewable and nonrenewable resources, and air/water recycling; and (3) Primary purpose of settlement (i.e., research, resources, commercialization). In addition, students will complete an application to settle their planet including cost/benefit analysis, sustainability of human habitation, and impact on local environment. The specific socioscientific issues (SSI) presented in this unit are:

1. How will extending human presence into and beyond the solar system affect society and culture on Earth?
2. What legal, ethical, and other value systems should govern human settlements and other activities in space?
3. Are space settlements colonies?
4. Do humans have rights to utilize or colonize extraterrestrial environments?

Connecting to *NGSS*

MS-ESS1-2. Develop and use a model to describe the role of gravity in the motions within galaxies and the solar system.

MS-ESS1-3. Analyze and interpret data to determine scale properties of objects in the solar system.

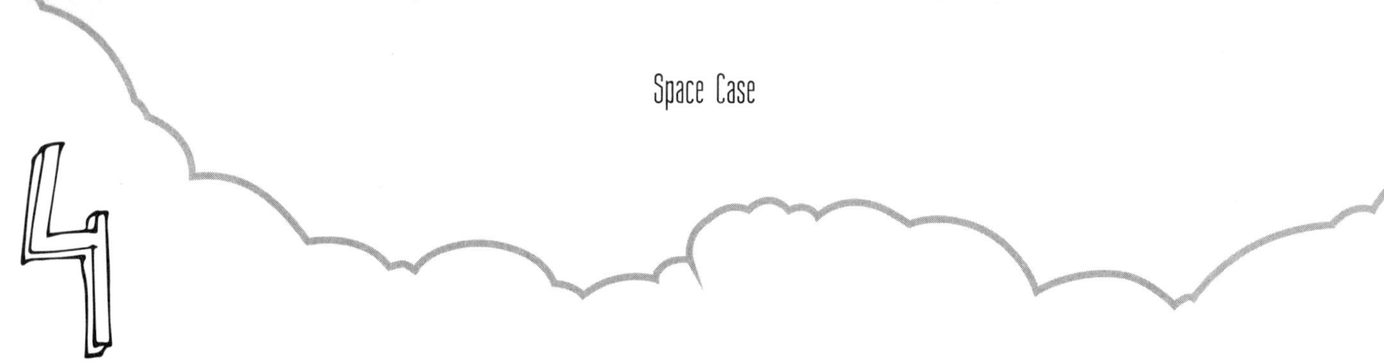

MS-ESS3-3. Apply scientific principles to design a method for monitoring and minimizing a human impact on the environment.

MS-ETS1-1. Define the criteria and constraints of a design problem with sufficient precision to ensure a successful solution, taking into account relevant scientific principles and potential impacts on people and the natural environment that may limit possible solutions.

MS-ETS1-2. Evaluate competing design solutions using a systematic process to determine how well they meet the criteria and constraints of the problem.

Accommodations for Students With Disabilities

Visual Disabilities: Use large-print formats for all reading material; allow student to read articles online so that magnification can be easily adjusted; allow student to become familiar with building materials in advance of the building day; diagrams can be made tactile using glue (and allowing it to dry) or Wikki Stix (wax covered strings).

Hearing Disabilities: Use closed captioning on video; introduce important vocabulary to student in advance so that he or she is familiar with it by the time it is used in a large group; be sure that only one student at a time speaks during class discussions or debates; give all instructions in verbal and written formats.

Learning Disabilities: Provide student with a graphic organizer to help her/him keep track of questions, evidence, and arguments; do a quick review at the beginning and end of each class to focus and refresh student memory; put a "Today's Focus" on the board so that the student has a daily goal; check reading level and adjust as needed.

Motor/Orthopedic Impairments: Be sure that the room setup allows students to move about easily; check that all activities are done at an appropriate height for a student who uses a wheelchair; during the building activity, try to provide materials that can be handled easily depending on the student's motor skills: for fine motor skills challenges, provide larger objects and adaptive writing implements and scissors; gloves with Velcro attached can be worn with additional Velcro placed on building items for easier pickup and manipulation.

Emotional Disabilities: This unit requires collaboration and consensus between teammates and also allows for debate, which can become quite heated. Assist students who have difficulties with compromising by mediating disputes before they erupt; remind teams that they have a common goal that can only be reached with everyone's input; provide students with communication and consensus-building techniques such as "interviewing" whereby students each provide their ideas, and the next student must repeat the ideas of the student before them and then ask two questions. This helps students focus on ideas rather than "being right." Allow for a "cool down" time if needed.

Resources and Research Materials

Billings, L. 2006. To the Moon, Mars, and beyond: Culture, law and ethics in space-faring societies. *Bulletin of Science, Technology and Society.* SETI Institute.

Livingston, D. 2000. Lunar ethics and space commercialization. Lunar Development Conference, Space Frontier Foundation.

Livingston, D. 2001. A code of ethics and standards for lunar development and outer space commerce. Lunar Development Conference, Space Frontier Foundation.

United Nations. 2002. Selected text from *United Nations Treaties and Principles on Outer Space: Part 1: UN Treaties.*

United Nations Educational, Scientific and Cultural Organization (UNESCO). 2009. Ethics of outer space: Cooperation with European Space Agency (ESA).

United Nations Educational, Scientific and Cultural Organization (UNESCO). 2009. Ethics of outer space: Starting point—Rereading "the ethics of space policy."

Video/DVD: *History Channel presents the universe: Colonizing space.*

Note: This unit is based on a project contributed by Tammy Modica.

Lesson 1

Who Owns Outer Space?

To the Teacher

In this introductory lesson, students will view a video that introduces them to the idea of extraterrestrial human settlements. They will then conduct research using various documents to begin to explore the socioscientific issues regarding planetary colonization and resource utilization. This lesson uses a Jigsaw strategy.

Objectives

Students will begin to consider the moral, ethical, scientific, and societal implications of this controversy and will attempt to formulate a sound decision on the issues of planetary settlements and resource utilization.

Time Needed

Three class periods

Materials

Video, *History Channel Presents the Universe: Colonizing Space* (available at: *http://shop.history.com/the-universe-colonizing-space-dvd/detail.php?p=254306&v=history_show_the-universe*)

The following articles:

- Billings, L. 2006. To the Moon, Mars, and beyond: Culture, law and ethics in space-faring societies. *Bulletin of Science, Technology and Society.* SETI Institute.
- Livingston, D. 2000. Lunar ethics and space commercialization. Lunar Development Conference, Space Frontier Foundation.

- Livingston, D. 2001. A code of ethics and standards for lunar development and outer space commerce. Lunar Development Conference, Space Frontier Foundation.
- United Nations. 2002. Selected text from *United Nations Treaties and Principles on Outer Space: Part 1: UN Treaties.*
- United Nations Educational, Scientific and Cultural Organization (UNESCO). 2009. Ethics of outer space: Cooperation with European Space Agency (ESA).
- United Nations Educational, Scientific and Cultural Organization (UNESCO). 2009. Ethics of outer space: Starting point—Rereading "the ethics of space policy."

Procedure

1. Ask students, "Who owns outer space?" After brief discussion, ask, "Why might humans be interested in "owning" space?" (resources, places to live, recreation) "What rules should be followed in space "properties," and who should determine them?" "What if there are other life forms in space … do they have rights?"
2. Explain to students that they have just begun to explore topics referred to as "space colonization" and "space law." During the next few weeks, they will become "experts" in these areas and will be simulating the process by which an extraterrestrial planet becomes settled. But before they can do that, they need to understand some basics about space colonization and space science.
3. Show the video *History Channel Presents the Universe: Colonizing Space* (Note: This video is 188 minutes, but can be viewed in sections to give an introduction to the topic. The video is presented in four segments: astrobiology, space travel, colonizing space, and living in space).
4. After the video, have students divide into teams of four. Give each team four of the articles listed in the Materials section. Have students divide up the articles so each has one. Explain to the students that they will work with others from the other teams who have their same article to become "experts" on their own article. In their expert groups, they will discuss the four questions below after reading their article. Finally, they will return to their original team and share what they learned and discussed with their expert team in order to more fully answer the questions.

5. Send students to their "expert" groups to read their articles and answer the following questions:

    - How will extending human presence into the solar system affect society and culture on Earth?
    - What legal, ethical, and other value systems should govern human settlements and other activities in space?
    - Are space settlements colonies?
    - Do humans have rights to utilize extraterrestrial resources and alter extraterrestrial environments?

6. After students have had time to work in their "expert" groups, they return to their original teams and report and discuss what they have learned. Teammates then work to develop more complete responses to the questions.

Closure

Ask, "Which questions do you feel were relatively easy to answer with the information you have?" "Which questions are more difficult?"

Assessment

Students will turn in their responses to the questions above.

4

Lesson 2

Space Players

To the Teacher

In this lesson students will assume the roles of various stakeholders involved in decision making regarding extraterrestrial settlements. They will then work with fellow students to further analyze the questions regarding space exploration/exploitation and will debate from the perspective of their assigned stakeholder position.

Objectives

- Students will review the information they have learned about extraterrestrial environments, requirements for life, and resource allocation.
- Students will produce valid arguments supported by scientific evidence.
- Students will learn about listening to each other's points of view in a critical manner

Time Needed

Two class periods

Materials

Articles from prior lesson, computers with internet access, student handout.

Procedure

1. Ask students, "Who is involved with making decisions about space colonization?" (United Nations, government agencies, scientific organizations, and so on)
2. Explain that today, students will be taking on the role of one of these "stakeholders" in order to debate questions about space colonization and resource usage in a simulated meeting at the United Nations to which various groups

have been invited in order to form new policies. Students who have similar roles will work together to research and prepare for the debate.

3. Students will be assigned to one of the following four groups for the role-play:

   A. United Nations: The governing body concerned about making sure that laws, policies, and treaties are applicable, enforced, and concerned for the benefit of all mankind.

   B. National Agencies: Agencies such as NASA, ESA, and JAXA that are committed to space exploration and scientific discovery. They are necessarily concerned about funding of their projects, which are threatened by budget cuts.

   C. Settlement Companies: Those that want to travel to another planet and settle it. They are the explorers, adventurers, scientists, and entrepreneurs. Many see economic advantages of utilizing extraterrestrial resources, while others want to participate in the establishment of "new worlds" that they can help guide.

   D. Special Interest Groups: Groups concerned about the environmental impact, ethical use, and allocation of space resources, financial burden of paying for projects, and protection of species (e.g., Greenpeace, Taxpayers Against Space Funding, Students Concerned About Alien Life, and so on).

4. Using the articles and any additional articles students have gathered from internet research, allow student groups to discuss the questions *from the perspective of their assigned role:*

   A. How will extending human presence into the solar system affect society and culture on Earth? Will it improve society by advancing technologies? Can it create more dissention on Earth?

   B. What legal, ethical, and other value systems should govern human settlements and other activities in space? Will settlements be democratic, communist, dictatorships, monarchies? Will settlements be public or private property? What laws should apply to settlements and who should determine them?

   C. Are space settlements colonies? Are they autonomous (settlements) or are they still governed by the sponsoring country (colony)? If it is a colony, what time delay will there be in governances due to extreme distances?

D. Do humans have the right to use extraterrestrial resources and alter extraterrestrial environments? What if there are no native life forms? What is the definition of "life form?" What constitutes a "sentient" life form? Should there be limits on types of factories or power sources that can be built on new worlds? If we're currently destroying our own planet (global climate change), should there be limits to how much "damage" can be done to another planet?

5. After groups have had an opportunity to answer their questions, convene a United Nations meeting to discuss these issues (with students representing their respective groups).

Closure

Ask, "Are we able to come to any consensus on any of the questions?" "What issues are the most contested?" "Can we reach a compromise?" "*Should* we reach a compromise?"

Assessment

Students are assessed on their research participation in their group as well as their participation in the meeting.

Scoring Rubric. United Nations Meeting Participation Sheet

(Place a check "√" in each box as task is completed)

Student Names	Completed Article Research	Participated in Stakeholder Group Discussions	Made Evidence-Based Contribution to United Nations Meeting	Developed and Expressed a Thoughtful Opinion on the Topic	Total (4 "√" Max)
United Nations					
National Agencies					
Settlement Companies					

Student Names	Completed Article Research	Participated in Stakeholder Group Discussions	Made Evidence-Based Contribution to United Nations Meeting	Developed and Expressed a Thoughtful Opinion on the Topic	Total (4 "√" Max)
Special Interest Groups					

United Nations Meeting Handout

The following groups have been invited to attend a meeting at the United Nations in order to form new policies regarding space settlements and extraterrestrial resource usage:

United Nations: The governing body concerned about making sure that laws, policies and treaties are applicable, enforced, and concerned for the benefit of all mankind.

National Agencies: Agencies such as NASA, ESA, JAXA, that are committed to space exploration and scientific discovery. They are necessarily concerned about funding of their projects, which are threatened by budget cuts.

Settlement Companies: Those that want to travel to another planet and settle it. They are the explorers, adventurers, scientists, and entrepreneurs. Many see economic advantages of utilizing extraterrestrial resources, while others want to participate in the establishment of "new worlds" that they can help guide.

Special Interest Groups: Groups concerned about the environmental impact, ethical use, and allocation of space resources, financial burden of paying for projects, protection of species. (e.g., Greenpeace, Taxpayers Against Space Funding, Students Concerned About Alien Life)

You have been invited to participate in that meeting as a member of one of the abovementioned groups. Using the articles you have read and any additional articles you can gather from internet research, work with fellow group members and discuss the questions *from the perspective of your assigned role:*

1. How will extending human presence into the solar system affect society and culture on Earth?
   - Will it improve society by advancing technologies?
   - Can it create more dissention on Earth?
2. What legal, ethical, and other value systems should govern human settlements and other activities in space?
   - Will settlements be democratic, communist, dictatorships, monarchies?
   - Will settlements be public or private property?
   - What laws should apply to settlements and who should determine them?
3. Are space settlements colonies?
   - Are they autonomous (settlements) or are they still governed by the sponsoring country (colony)?
   - If it is a colony, what time delay will there be in governances due to extreme distances?
4. Do humans have rights to use extraterrestrial resources and alter extraterrestrial environments?
   - What if there are no native life forms?
   - What is the definition of "life form?"
   - Should there be limits on types of factories or power sources that can be built on new worlds?

Lesson 3
Planetary Particulars

To the Teacher

During this lesson, students will perform extensive research on the relationships between planets and other astronomical bodies relative to the universe, including distance, size and composition in order to develop a detailed diorama, poster, and application for settlement for their extraterrestrial planet.

Objectives

- Students will explore the structure and diversity of planetary systems.
- Students will develop detailed models of planetary features.
- Students will investigate possible resource uses for their settlement.
- Students will utilize cooperative skills such as consensus-building and compromise to develop an application for a planetary settlement.

Time Needed

Two to three class periods

Materials

Student handouts ("Alien Earth" and "Application to Settle Extraterrestrial Planet"), one poster board per team, materials to make a planet diorama (clay, paper, beads, string, LEGOs, and such), computers with internet access

Procedure

1. Ask students, "If you could begin a settlement anywhere in the universe, what planetary features would you look for?"
2. "Today you are going to have the opportunity to work with teammates to establish your own extraterrestrial settlement. In order to do so, you will need to consider various aspects of the

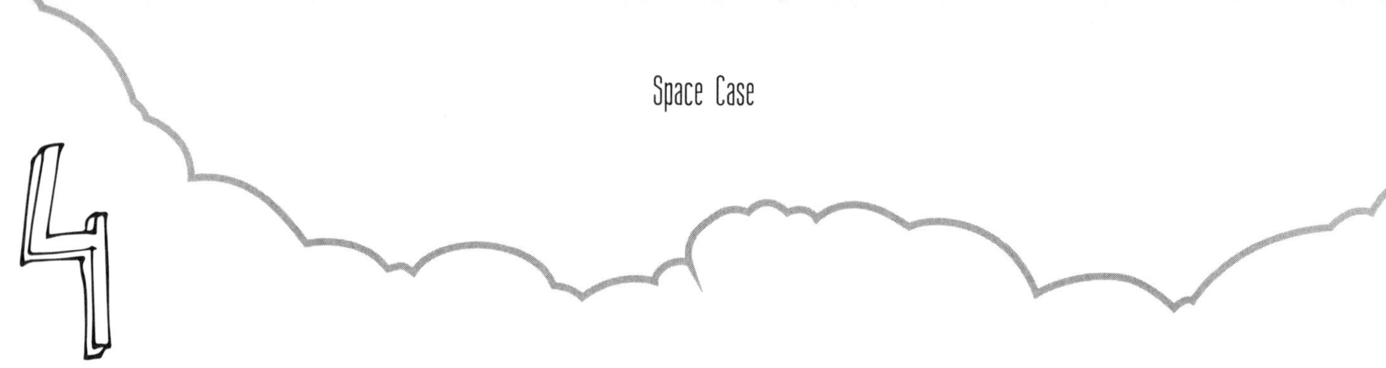

planet's geological features, atmosphere, and resources. You are challenged to create a diorama and poster about your settlement. You will also complete a detailed application for settlement to be submitted to the United Nations for consideration. The specifics of your project are on these two handouts."

3. Allow students time to work on their dioramas, posters, and applications according to the specifics in the handouts.

Closure

Inform students that they will be presenting their applications to the United Nations for a vote of approval or denial of permission to settle during the next class.

Assessment

Students will be assessed per the scoring information on the posterboard and diorama handout below.

"Alien Earth"

Planet Poster Board and 3D Diorama of Your Planet's Surface

Objective
To explore the structure and diversity of planetary systems.

Materials
Poster board and materials for 3D diorama of your planet's surface.

Procedure
Your settlement group (2–3 people) is to create and depict an extraterrestrial planet, which you will then colonize.

***Diorama* (30 pts.):**
Create a 3-D model of the surface of your planet and its colony. (10 pts. each)

1. Depict all geological features. (Does it have steep or rolling mountains? Are there any rivers, lakes, waterfalls, or oceans? Are there volcanoes and are they active? Are there endless sand dunes? Is there a diamond waterfall?)
2. Depict some of the colony present on your planet. (Is there a big city? A small colony? A luxury resort casino and hotel? A mining compound? Ancient ruins?)
3. Make sure to add color, details, and labels to your diorama.

***Maximum size** allowed is 2' × 1' × 1'. Projects that exceed this size limit will lose points.

***Poster* (70 pts.):**
Describe your planet in detail using words and pictures. (10 pts. each = 50 pts. total)

Each item below should have its own (5 pts) identifying paragraph (4–8 sentences) and (5 pts) image/graphic.

1. How many (#) moons does your planet have, what are their sizes, and how close or far away are they?
2. What is your planet's distance from its primary star (use our solar system orbits as

comparison: "Closer than Mercury's orbit"), what is your planet's relative size (same size as Earth, twice as large as Jupiter?), and what is your planet's acceleration of gravity with respect to Earth's? (a third of Earth's, same, 3 times as much?)

3. What type of star system does your planet have? (regular, red giant, binary, pulsar, white dwarf)
4. Does your planet have any geological activity? (Minor/major earthquakes; volcanoes; hurricanes)
5. What are the main elements present on your planet (is it a "water world," a "volcano world," a "diamond world," a "uranium world," an "Earth-like world") and what is the atmosphere like? (thin/thick carbon dioxide, nitrogen/oxygen, sulfuric acid, methane, cloudy with lots of storms, clear with hardly any clouds, heavy/no rains) [Hint: What elements are in your atmosphere will help determine the color of your sky. You can research this kind of information!]

***Can You Support Life?* (5 pts. each = 20 pts. total)**

1. What type of habitable living areas do you have? (houses, pods, aerial, underground, orbital)
2. What power source is there? (solar, water, wind, nuclear, geothermal, something new)
3. How is the water and air supply recycled? (water tower, bio-dome, plant-life)
4. What types of scientific laboratories are there and what are they studying? (astronomy: studying your star system; biology: studying new forms of life or how life grows on your planet; chemistry: studying chemical reactions and new elements/ molecules; geology: how was the planet formed)

Application to Settle Extraterrestrial Planet

United Nations Office for Outer Space Affairs (UNOOSA)

APPLICATION FORM

I hereby apply to participate in the United Nations Settlement of Extraterrestrial Planets, recognizing the common interest of all mankind in the progress of the exploration and use of outer space for peaceful purposes, desiring to contribute to broad international co-operation in the scientific as well as in the legal aspects of exploration and use of outer space and believing that such co-operation will contribute to the development of mutual understanding and to the strengthening of friendly relations between nations and peoples. Taking into consideration the resolutions 1721 (XVI) of 20 December 1961 and 1802 (XVII) of 14 December 1962, adopted unanimously by the States Members of the United Nations, I solemnly declare that my settlement, in the exploration and use of outer space, should be guided by the principles adopted by the General Assembly of the United Nations 1962.

A. Personal Data

1. Full Name: ____________________
2. Organization: ____________________
3. Primary Reason for Settlement: ____________________

B. Sustainability of Human Habitation

1. Please list all resources being brought to the planet by your crew and spacecraft:

____

____

____

____

2. Please list any resources from the planet that you intend on using and their effect on the local community:

____

____

____

____

3. Please describe, in detail, how your settlement intends to recycle breathable air and drinkable water for your community and its effect on the local community:

____

____

____

____

____

____

____

____

____

C. Cost/Benefit Analysis

1. Size of crew: ____

2. Duration of flight from Earth to Planet: ______________________________

3. Please indicate if your company is able to fund this settlement, covered either by your sponsoring agency/organization, or another international, regional, or national organization: ____

____

____

____

4. Please describe how long your settlement will take to become fully operational:

___

___

___

5. Please provide detailed information on the benefit of creating your settlement on this planet and how this will contribute to the common interest of all mankind in the progress of the exploration and use of outer space for peaceful purposes:

___

___

___

___

D. Would you be prepared to make a presentation on the purpose of your settlement and the nature of your chosen planet as well as the status of space law and policy in your company?

Yes (   )   No (   )

E. Please provide a brief summary of any developments in your company relating to space law, policy, and settlements:

___

___

___

___

___

___

___

___

F. Applicant's Signature

___

(Signature of Applicant) (Date)

Lesson 4

United Nations Vote

To the Teacher

In this final lesson, students will present their planet settlement proposals, including their dioramas, posters, and applications to the "United Nations" for a vote on approval or denial of permission for their planetary settlement.

Objectives

- Students will review the requirements for life by considering the atmosphere and resources on their planet.
- Students will become familiar with the physical features and scale involved in planetary exploration.
- Students will consider the scientific, economic, ethical/moral, and social implications of extraterrestrial planetary settlements.
- Students will utilize scientific evidence to develop arguments for their planetary settlement.

Time Needed

One class period

Materials

Students' dioramas, posters, and planetary settlement applications.

Procedure

1. Student settlement groups are given five minutes to present their "case" for their planetary settlement using their diorama, poster, and application. Time should be allowed for questioning.

2. United Nations members (classmates and teacher) should consider the following in determining their vote: (a) Did the settlement company present a scientifically valid assessment of its planet? (b) Did the settlement company consider all of the factors involved in the "Alien Earth" sheet? (c) Did their diorama and poster present a convincing case for the settlement's existence?
3. Students can vote on applications and discuss.

Closure

Ask, "What aspects of extraterrestrial settlements do you feel are fairly well settled?" "What aspects do you feel still require much negotiation? "How did you knowledge of science help you to understand the issues presented in this unit? "What considerations, besides scientific facts/evidence, are important in evaluating extraterrestrial colonization and/or resource usage?"

Assessment

Students are assessed on their projects as well as a reflection page on the above "Closure" questions.

Scoring Rubric

(*Note:* Scoring for the poster board and diorama is included in the "Alien Earth" student handout.)

Did the settlement company present a scientifically valid assessment of its planet?	____ 20 pts.
Did the settlement company consider all of the factors included in the "Alien Earth" Sheet?	____100 pts.
Did the settlement company complete the settlement application in a clear, accurate, and compelling manner?	____ 50 pts.
Did the student write a thoughtful reflection page on the "Closure" questions?	____ 20 pts.
Do all materials contain proper grammar and structure?	____ 10 pts.

Total Pts. _____ /200

Unit 5

A Fair Shot?

Should Gardasil vaccines be mandatory for all 11–17-year-olds?

Unit Overview

During this unit, students will gain a rich understanding of the human immune system by studying the interactions between immunity and vaccinations, relationships between certain viruses and cancer, and the mechanisms of allergic reactions. By embedding this content into a controversial question about a mandatory vaccine, students gain insight into the personal, societal, and economic impacts of scientific innovations and learn the importance of informed participation in scientific policy debates.

Key Science Concepts

Human Immune Function, Allergic Response, Mechanisms of Antibiotics, Relationship of Genetics and Viruses tto Cancer, Transmission and Prevention of STDs

Ethical Issues

Personal Freedom, Privacy Rights, Paternalism, Issues of Minors, Legal Consent, Economics of Medicine

Science Skills

Organizing Information, Observing, Understanding Cause and Effect, Communicating Results, Interpreting and Representing Data, Forming Arguments From Evidence

Grade Levels

High School Advanced Biology or Anatomy; can be adjusted for General Biology

Time Needed

The unit is comprised of five lessons of approximately 50 minutes each. Time can be adjusted depending on student content background.

Lesson Sequence

Lesson 1. Introduction: Should the Gardasil Vaccination Be Required for All 11–17 Year Olds? An Immune System Research Project

Lesson 2. The Biology of Cancer

Lesson 3. It Can't Happen to Me! Sexually Transmitted Infection/Diseases

Lesson 4. The Art of Argument: Research and Debate

Lesson 5. Rethinking Positions and Relating the Gardasil Debate to the Nature of Science (NOS)

Background on the Issue

The goal of this unit is to promote respectful discourse among students while learning about vaccinations, the immune system, and cancer, as well as a current topic of global debate. The controversial aspect of this unit deals with the requirement of the Gardasil vaccine. The Gardasil vaccine protects against 4 of the 30–40 HPV (Human Papilloma Virus) types. Two of the four HPV types it protects against cause 90% of genital warts cases in males and females. The other two HPV types cause 75% of cervical, 70% of vaginal, and 50% of vulvar cancer cases in females (*www.Gardasil.com*). Gardasil is also effective in protecting against HPV-related anal and penile cancer in males. Several states have implemented requirements of vaccination for girls, with current pending legislation in many states for all students. Although the vaccine is highly efficacious against HPV, several groups have opposed mandatory legislation based on fears about vaccines in general, as well as privacy and personal autonomy rights. Equity and economic issues have also emerged given the fact that the vaccine is effective against genital warts and cancers for both males and females, but legislation has focused primarily on females due in part to the fact that penile and anal cancers in males are less common than the cancers in females, thus making the male vaccine less cost-effective. With that said, the socioscientific issue (SSI) in this unit is: Should the vaccine be required for *all* males and females ages 11–17?

The methods that will be used to explore this SSI will include, but are not limited to, the use of technology, collaborative and cooperative groups, evoking ethos in order to engage students, videos, student presentations (students teaching students), inquiry-based microscope activities, discussions, student-based research, and student-led debates. This issue is a most timely and relevant one for young adults and is sure to inspire curiosity about the immune system, vaccinations, and the numerous considerations involved in rendering health care policies.

Connecting to *NGSS*

HS-LS1-1. Construct an explanation based on evidence for how the structure of DNA determines the structure of proteins which carry out the essential functions of life through systems of specialized cells.

HS-LS1-2. Develop and use a model to illustrate the hierarchical organization of interacting systems that provide specific functions within multicellular organisms.

HS-LS1-4. Use a model to illustrate the role of cellular division (mitosis) and differentiation in producing and maintaining complex organisms.

HS-LS3-2. Make and defend a claim based on evidence that inheritable genetic variations may result from: (1) new genetic combinations through meiosis, (2) viable errors occurring during replication, and/or (3) mutations caused by environmental factors.

HS-ETS1-3. Evaluate a solution to a complex real-world problem based on prioritized criteria and trade-offs that account for a range of constraints, including cost, safety, reliability, and aesthetics, as well as possible social, cultural, and environmental impacts.

Accommodations for Students With Disabilities

Visual Impairments: This unit involves extensive research from print and online sources. Be sure to provide large-print handouts for low-vision students, use closed-circuit television for projection, or allow students to do all tasks on large computer monitors with screen magnification or enhancement software; blind students will require screen reading software or printed materials translated to Braille; if adaptive microscopes are not available, tactile models of cells can be created out of clay, puffy paints, or Wikki Stix; during the Lesson 3 simulation activity, have

all students describe their results verbally as they are appearing (i.e., "My liquid stayed clear!") so that the visually impaired student(s) can fully participate.

Hearing Impairments: Give instructions in written and verbal formats; be sure to face the student when you are speaking to him/her (or the class); remind teammates and classmates to face student when speaking; during debates or class discussions, have student speakers give a signal (such as a slight hand wave) to alert student to the source of the sound; ask student to alert you to any confusion she or he is having with vocabulary word pairs that might sound alike or appear alike to a lip reading student (such as HPV and STD, B cells and T cells, and so on).

Learning Disabilities: Provide graphic organizers to help students keep track of the facts they learn and the arguments those facts support; use "wait time" for all students; provide students with highlighters, sticky notes, or other aids for organizing notes and readings; display group "Journeys Through the Immune System" projects in the classroom so that they can act as visual cues for important information during the remainder of the unit.

Motor/Orthopedic Impairments: Allow student to utilize alternative note-taking devices including voice-to-text software if writing is not feasible; microscope knobs can be enlarged by placing plastic wrap around the knob (for protection) and then covering the plastic with clay, making the microscope easier to manipulate; during the Lesson 3 simulation, make sure that the student has ample space for "mingling;" the cup can be attached to a wheelchair arm using Velcro so that the student can move around freely and mix the liquids as needed.

Emotional Disabilities: This unit involves several group activities. Remind all students that a large part of the process in working through meaningful socioscientific issues is practicing cooperative skills such as consensus-building, collaboration, compromise, and communication; allow students "time out" if discussions become too heated or behavior becomes inappropriate; allow students to express frustrations or challenges in appropriate ways, such as journaling, drawing, one-on-one conversations, or online discussion boards with you.

Resources for Teachers

- Gardasil: *www.gardasil.com*
- National Institute of Allergy and Infectious Diseases: *www.niaid.nih.gov*
- Public Broadcasting Service: *www.pbs.org*
- National Cancer Institute: *www.cancer.gov*

- PBS Learning Media: *www.teachersdomain.org*

Note: This unit is based on material submitted by Ashley Schumacher, Michael Caponero, Brian Brooks, and Bryan Kelly.

Lesson 1

Introduction

Should the Gardasil Vaccination Be Required for All 11–17-Year-Olds? An Immune System Research Project

Topic: Vaccines
Go to: *www.scilinks.org*
Code: SSI008

To the Teacher

In this initial lesson, students are confronted with a challenging issue that will evoke emotional reactions, but requires more scientific background than students will likely have to make a thoughtful, reasoned decision. After a brief introduction and discussion of the question, students will perform research on the human immune system and try to develop 10 evidence-based arguments for or against the issue. Students will discover that they need more information to complete this task. Students will then develop a project to communicate their understandings of the human immune system.

Objectives

Part 1

- Students will begin to consider the moral, ethical, scientific, and societal implications of this controversy and will attempt to formulate a sound decision on the issue.

Part 2

- Students will be assigned a creative project that will actively engage the students in learning about the immune system. Project Science Content objectives:

    1. The students will describe the function of the immune system.

2. The students will explain how the skin functions as a defense against disease.
3. The students will distinguish between a specific and nonspecific response.
4. The students will describe the actions of B cells and T cells in an immune response.
5. The students will describe the relationship between vaccination and immunity.
6. The students will describe what happens in an allergic response.
7. The students will describe at least one immune disorder.
8. The students will explain (diagram) the antigen-antibody reaction.

Time Needed

One period for Part 1; Part 2 can be done in class over 2–3 periods or as a group homework assignment using fewer class periods.

Materials

Part 1

- Student position statement handout

Part 2

- Booklet: Understanding the Immune System: How It Works. (Download here: *www.niaid.nih.gov/topics/immuneSystem/Documents/theimmunesystem.pdf*)
- Computers with internet connection

Procedure

Part 1

1. Begin by asking the students if they would take a vaccination that would almost certainly prevent them from getting cancer. (Most students will say "yes.") Then inform them that although there is a "high chance" that the

vaccine will prevent them from getting cancer, there is also a "small chance" that it may also cause them to become ill and even die in a different way. The students who answered "yes" might reconsider their answers, or they may find that the "high chance" outweighs the "small chance." Provide time to hear the opinions of the students.

2. Present students with some general background information regarding the Gardasil vaccination (i.e., helps protect against certain viruses that can cause cancer, allergic reaction is possible, does not protect against all viruses, effective in both males and females who have not gotten exposed to virus).
3. Explain that several states are considering mandatory Gardasil vaccination program for all students ages 11–17. Allow the students to provide feedback regarding this scenario.
4. Distribute a copy of the "Student Position Paper" on which students will write their position on the issue for or against the mandatory vaccination program.
5. Ask the students if they are able to come up with the 10 supporting statements for their positions. It is assumed that most students were unable to meet this requirement; discuss statements to evaluate whether they are able to support their arguments with facts, or simply opinions.
6. Ask, "Do you have enough information to make an informed decision?" "If not, what other types of information do you need?"

Part 2

1. Assign the students to groups of 3 or 4.
2. Challenge groups to create a poster/PowerPoint/pamphlet/performance describing a "Journey Through the Immune System," which will be presented to the class. The groups will be given a handout detailing the assignment and the required objectives. In order to prepare, groups will be given the "Understanding the Immune System" booklet and computers with internet access.
3. Allow two periods for the group research, development, and presenting of team projects.
4. Allow teams approximately 10 minutes to present their project followed by 5 minutes of questions and answers.

5. Ask students to revisit their argument sheets regarding the Gardasil question. Do they have enough information yet? If not, what is lacking? (understanding of cancer, relationship between virus and cancer, the nature of STDs)

Closure

Inform students that over the next several lessons, they will be learning more about the immune system, vaccinations, and cancer in order to make informed decisions about the Gardasil vaccine.

Assessment (Homework)

Students will watch the following video: *www.teachersdomain.org/asset/frntc10_vid_vaccines.*

The students will answer the following questions:

1. Through a published schedule and set of guidelines, the Centers for Disease Control and Prevention (CDC) and public health officials recommend that every child receive certain vaccinations by age 6. What are the benefits of this recommendation to public health officials, to the community, and to other children?
2. Some parents and health care professionals question the CDC's recommendations and decide not to vaccinate their children, while others, like Jennifer Margulis, choose to vaccinate their children along an alternative schedule. How might her decision affect both her own children and others?
3. In what ways is vaccination different from other types of personal health decisions? Who should be involved in deciding whether children receive a specific vaccine?
4. Should the government have the right to compel vaccination? Should parents have the right to refuse it?

Name: ______________________________ Date: ______________

Student Position Statement

Should Gardasil be made a mandatory vaccination for all students ages 11–17? *(Check one)*

Yes (   ) No (   )

Provide at least 10 reasons or statements of evidence to support your claim:

1. ______________________________

____

2. ______________________________

____

3. ______________________________

____

4. ______________________________

____

5. ______________________________

____

6. ______________________________

____

7. ______________________________

____

8. ______________________________

____

9. ______________________________

____

10. ______________________________

____

"Journey Through the Immune System" Project Requirements

Using your prior knowledge, the "Understanding the Immune System" booklet (*www.niaid.nih.gov/topics/immuneSystem/Documents/theimmunesystem.pdf*) and valid internet sources, come up with a 10-minute presentation detailing a "journey" through the immune system that covers the following objectives:

1. Describe the function of the immune system.
2. Explain how the skin functions as a defense against disease.
3. Distinguish between a specific and nonspecific response.
4. Describe the actions of B cells and T cells in an immune response.
5. Describe the relationship between vaccination and immunity.
6. Describe what happens in an allergic response.
7. Describe at least one immune disorder.
8. Explain (diagram) the antigen-antibody reaction.

All presentations should include the following vocabulary:

Immunology, antigen, antibody, lymphocyte, leukocyte, thymus gland, bone marrow, B-cell, T-cell, macrophage, vaccine, antibiotic, inflammatory response, immune response, antihistamine, autoimmune disease, fever, helper T cell, pathogen, killer T cells, interferon

Be creative with your presentations! You may make them entertaining but they must include valid science knowledge and adhere to the specifications above.

Some suggestions: Posters, PowerPoint presentations, ppamphlets, performances. Add graphics, cartoons, diagrams, or props that help convey the information.

Journey Through the Immune System Scoring Rubric

Did the project convey the required information?	____ 30 pts.
Did the project include the required vocabulary?	____ 20 pts.
Was the project's information accurate?	____ 20 pts.
Did the members of the group participate equally?	____ 10 pts.
Was the project presented clearly?	____ 10 pts.
Was the project creative?	____ 10 pts

Total Pts.____________/ 100

Lesson 2

The Biology of Cancer

To the Teacher

Since the Gardasil vaccine protects against the HPV viruses that cause many vaginal, cervical, and vulvar cancers in women, as well as penile and anal cancer in men, it is essential that students understand the nature of cancer, the relationship between cancer and cellular reproduction, and the genetic mechanisms involved in cancer growth in order to understand the potential impact of the vaccine. In this lesson, students will compare normal and cancerous biopsy cells under the microscope, view an interactive website on cancer, and view a cancer PowerPoint handout to gain an understanding of the nature of cancer.

Objectives

- Students will apply their prior knowledge of cell reproduction.
- Students will learn about the cell biology of cancer based on cell reproduction.
- Students will learn about benign and malignant tumor cells.
- Students will learn there are many causes of cancer.
- Students will learn cancer stops the normal regulation of cell growth.
- Students will learn cancer develops as genetic damage of cells.

Time Needed

One class period

Materials

- Slides of biopsies positive for cancer
- Slides of healthy cells of same area as cancer biopsy slides

- Computers with internet access for groups of two
- Cancer PowerPoint handout (This can be found in the Resource section.)
- Links to videos (will be linked into the procedure at the accurate time)

Procedure

1. Students will be given two slides (one slide of healthy cells and one of cancer cells)
2. Students will be given no instructions other than "You are getting two slides of cells from the same area of the body."
3. After students have been given time to observe and draw the two different types of cells they should come to the conclusion that there is something different with the two slides of cells. The teacher should lead them with a series of questions to the idea that the one of the slides shows cancer.
4. Students will then, in groups of two, go to the computers and view the interactive website: *www.pbs.org/wgbh/nova/cancer/grow_flash.html*.
5. Students will work together to answer the "Questions to Ponder" on the PowerPoint handout.

Closure

Ask, "How does our study of cancer relate to the question of whether the Gardasil vaccine should be mandated?"

Assessment

Students will be assessed on cell slide observations and responses to "Questions to Ponder" PowerPoint handout.

Cancer PowerPoint Questions to Ponder

(from *www.cancer.gov/cancertopics/understandingcancer/cancer/page1)*

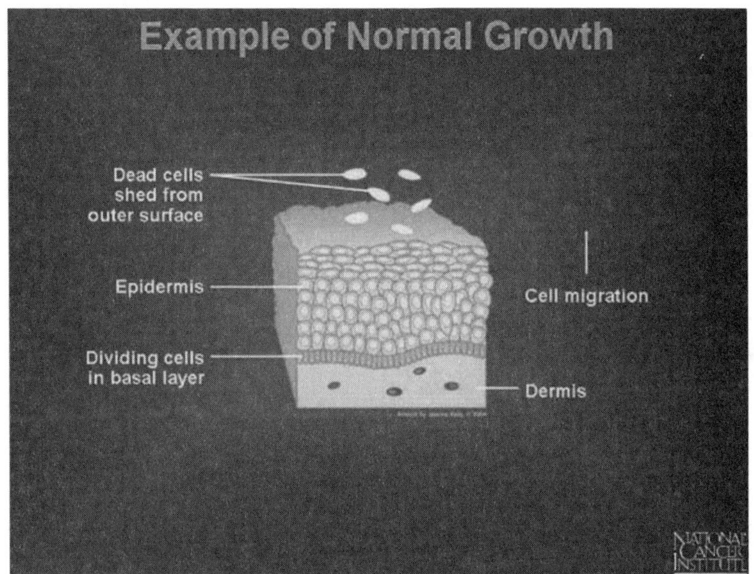

Questions to Ponder

What within the cell has mutated to cause the mutated cell?

What might you expect to happen if the uncontrolled growth does not become controlled?

How might the uncontrolled growth be controlled?

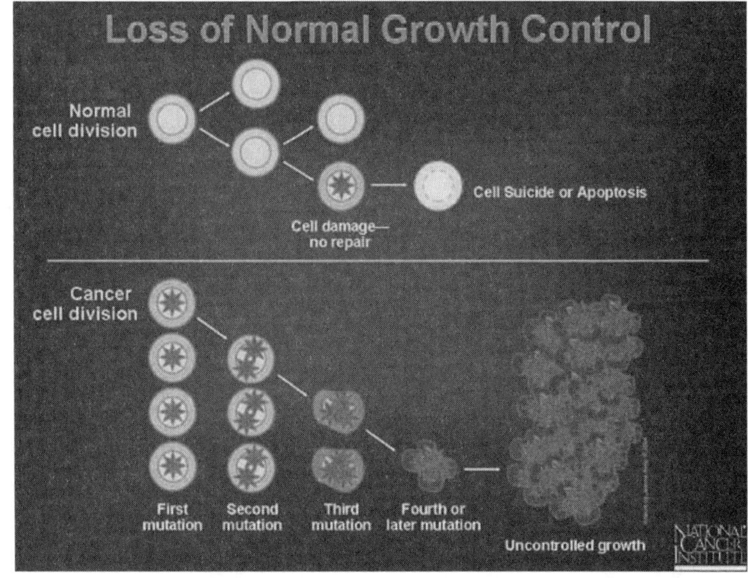

Questions to Ponder

What causes cells to die?

Is cell death a good thing?

What happens if no cells die?

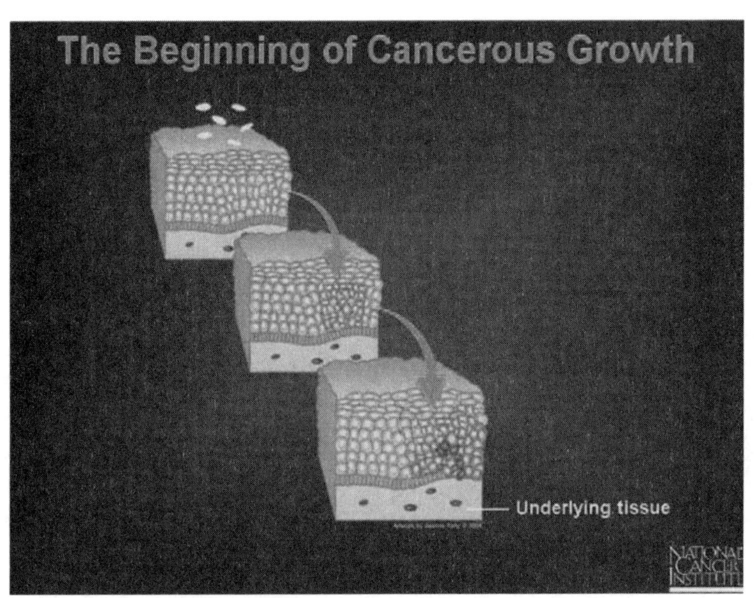

Questions to Ponder

Are cancerous growths green?

Do you think the cancerous groups from the first to the second picture and the second to the third picture happens in one cell division?

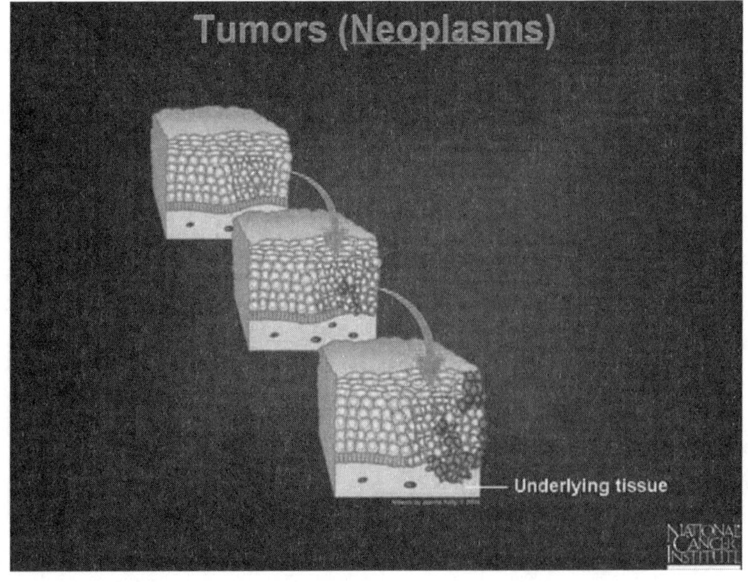

Questions to Ponder

What is a tumor (aka a neoplasm)?

Why is the green in picture three moving into the underlying tissue?

What does the green moving to the underlying tissue in picture three signify?

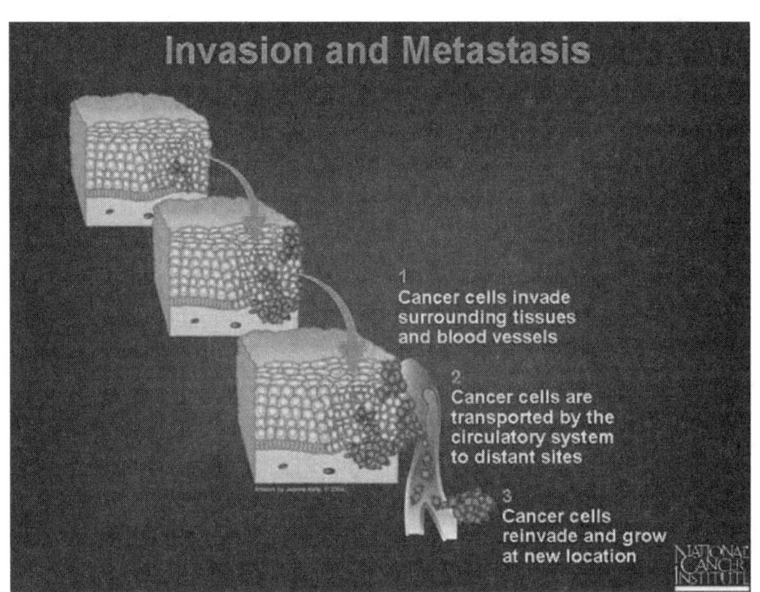

Questions to Ponder

What do the cancer cells invade?

What does metastasis mean?

What do you expect to happen after the cancer cells have started to grow in the new location?

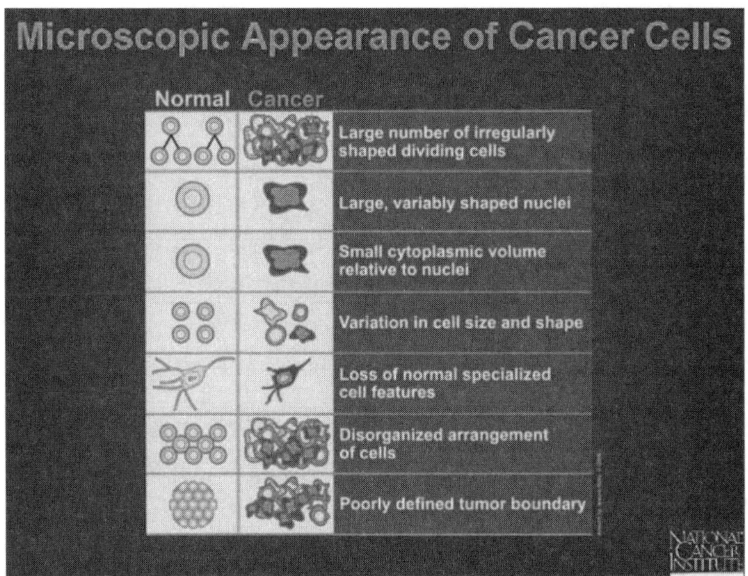

Questions to Ponder

Do these pictures look like anything we saw in class today?

Which column do you think is the cancerous cells? Which column represents the normal cells?

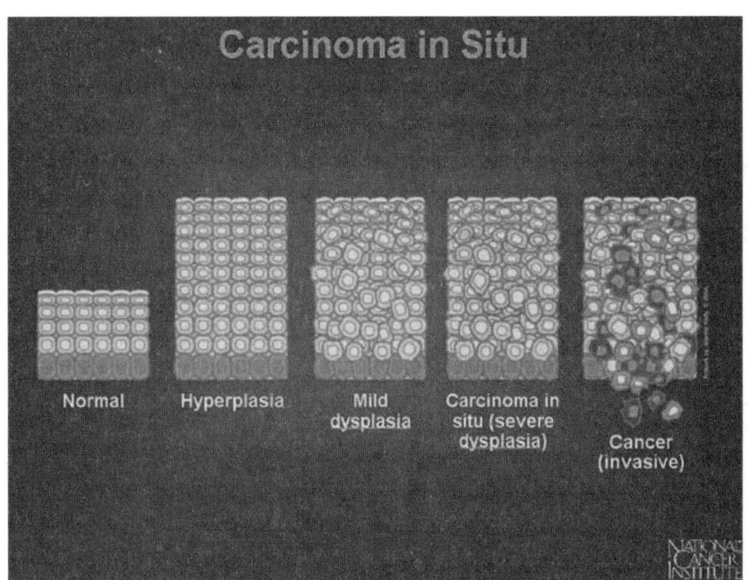

Questions to Ponder

What do you think hyperplasia and dysplasia mean (without looking it up, try to use the pictures to figure it out)

What causes the cancer (invasive) to leave the area of cells?

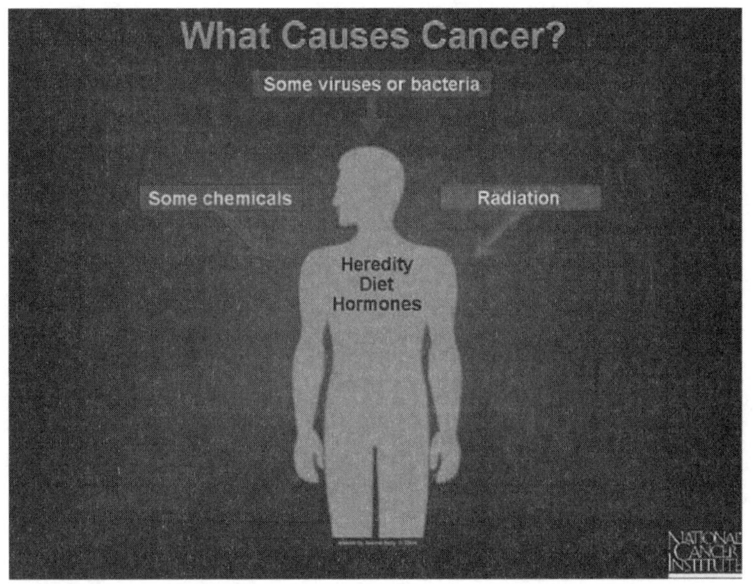

Questions to Ponder

Using your prior knowledge, name three causes of cancer (preferably one of each type but if you can't come up with one of each you can state multiple causes of the same type)

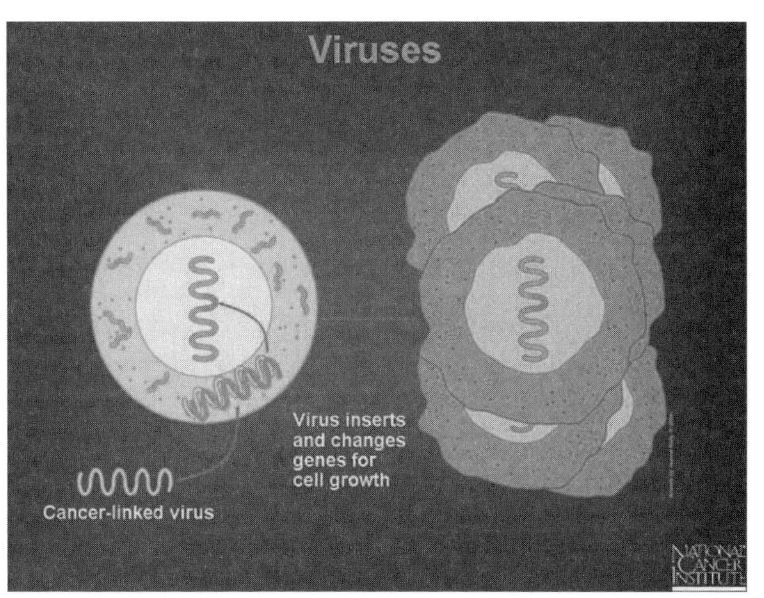

Questions to Ponder

What is a kind of virus linked with cancer we have discussed in this unit?

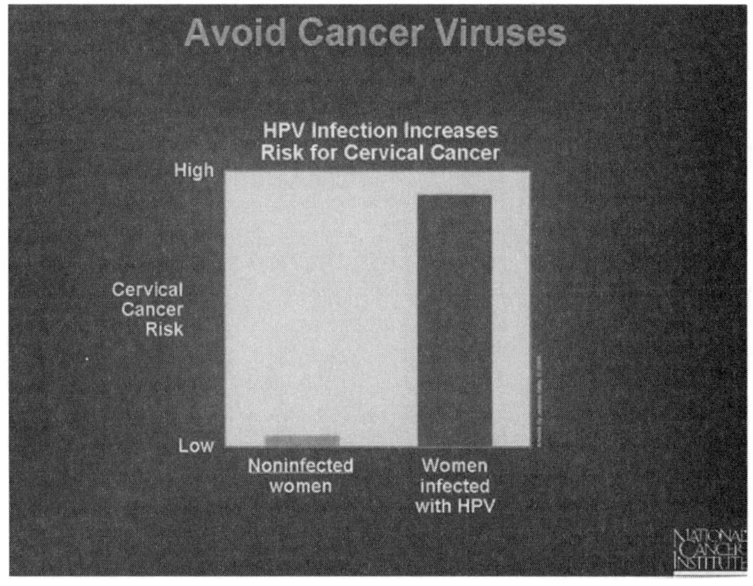

Questions to Ponder

What do you notice from this slide?

Does this slide encourage you to get the Gardasil vaccination?

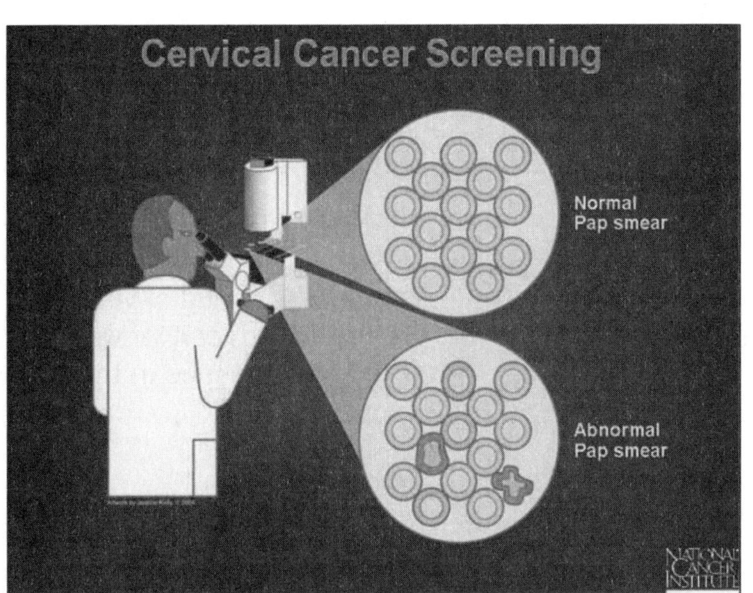

Questions to Ponder

Which Pap smear resembles which slide we looked at during class today?

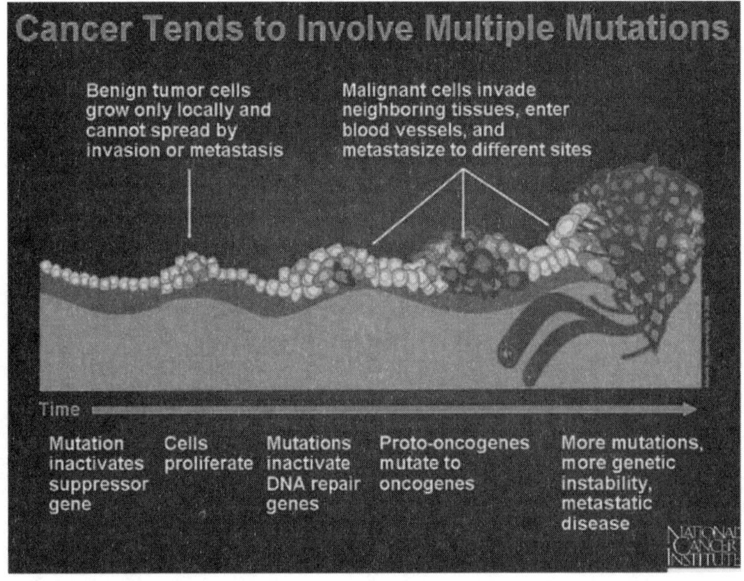

Questions to Ponder

What does benign mean?

What does benign mean in regards to a tumor?

Do benign tumor cells turn into malignant tumor cells?

Lesson 3

It Can't Happen to Me!

Sexually Transmitted Infections and Diseases

To the Teacher

In this powerful simulation activity, students will model the spread of STDs through populations. By using water and indicators to mimic the sharing of bodily fluids tainted with infection, students will experience the vulnerability of being a recipient, the responsibility involved in potentially spreading a disease to others, and the swiftness with which these diseases can travel. Students will then engage in research to debunk common misconceptions about these diseases.

Objectives

- Students will learn how STDs can travel undetected through a population.
- Students will learn about common misconceptions involving STDs.
- Students will learn about valid science vs. pseudoscience.

Time Needed

One class period

Materials

Part 1: Simulation Activity

- 1 clear plastic cup for each student
- Phenolphthalein
- Sodium hydroxide (tablet)

- Syringe or pipette
- Music source such as computer, radio, CD player (optional)

Part 2: Myth-Busting Research

- STDs Fact or Fiction sheet
- Computers with internet access

Preparation for Activity

Before students arrive to class, fill *one plastic cup per student* a little less than half-full with water. Numbers may be written on the bottom of each cup to complete a transmission diagram afterward; however, this is not necessary. Add one NaOH tablet to ONE of the half-filled cups and allow it to dissolve. It will look and smell like water. The person who receives this cup will be the source of the infection!

Procedure

Part 1

1. As students enter the class, have each student select a plastic cup.
2. After students are seated, explain to them that this liquid represents their immune systems. Ask, "Does your fluid look suspicious?" Does it smell funny?" "Does it look different than your neighbors'?"
3. Instruct them that they are to mingle as if they were at a party while music is playing. Once the music stops they will share the fluids in their cups with the student closest to them by completing the following procedure. (Model this.) They will pour the contents of their cup into the other student's cup. Now the other student's cup will be full. The second student will then pour all of the contents of his cup into the first student's cup. Finally, the first student will fill the second student's cup half-full and be seated.
4. Repeat the music, mingle, and mix procedure several times (three times is sufficient but more can be done).
5. Once you stop the music for the last time, have students return to their seats and ask the following questions: "Does your fluid look 'infected' now?" "Does anybody feel like a 'risk-taker'?" "Knowing that one of you is

infected, and assuming that you are the one, how do you feel about having shared your infection with others?"

6. Begin testing for STDs by the following process: Drop one drop of phenolphthalein into a student's cup. If it turns pink, then he or she has been infected. If it stays clear, he or she is healthy.

Closure for Part 1

1. Ask, "Do we know which person was first infected?" (No, but the first transmission might be able to be determined by how deep a color their liquid turned.)
2. Ask, "How does this activity mimic the transmission of communicable disease?"
    - A person who is infectious frequently does not exhibit any outward symptoms at the beginning of an illness so it is not always possible to tell who is ill by sight;
    - Communicable diseases are easily spread through a variety of methods including simply touching an area that an infectious person has recently touched. This activity mimics the rapidity with which a communicable disease can move through a population.

Extension for Part 1

This activity can be repeated by having students "mix" with half of the original number of partners. Students will see the vast reduction in the number of infections.

Note: This activity was adapted from Montana Office of Public Instruction. (n.d.). *K–12 teaching strategies for the prevention of unintended pregnancy and HIV/AIDS/STD.* (HIV/AIDS/STD Program). Helena, MT: Montana Government Printing Office.

Procedure

Part 2

1. Students will be divided into groups of four students. Each group will be given four statements from the "Fact or Fiction" list.
2. The groups will have 10 minutes to research the statements to determine if they are fact or fiction.

3. Students will come back together to present their conclusions to the class.
4. Review answers for the fact or fiction sheet.

Assessment

Students write a reflection paper about their experience with the simulation and the "fact or fiction" discussion. Papers should include thoughts about knowledge gained, insights into their own behaviors or feelings, and any questions they still may have.

Scoring Rubric

1 pt.	2 pts.	3 pts.	4 pts.	5 pts.
Reflection fails to convey information on knowledge gained or reflect on student's own behaviors or feelings, or remaining questions.	Reflection includes some insights into student's own behaviors or feelings, but fails to include any information on knowledge gained or questions remaining.	Reflection conveys information on knowledge gained, but fails to reflect on student's own behaviors or feelings.	Reflection conveys information on knowledge gained, insights into student's own behaviors or feelings, and any remaining questions, but does so in an unclear or confused manner.	Reflection conveys information on knowledge gained, insights into student's own behaviors or feelings, and any remaining questions in a clear and thoughtful manner.

STDs: FACT OR FICTION

STD facts

1. Birth control pills do not prevent sexually transmitted diseases from being contracted.
2. Over 95% of all STDs are contracted through sexual intercourse.
3. Most STDs can be treated.
4. There is no cure for herpes or AIDS.
5. STDs cannot be transmitted by touching doorknobs, drinking fountains, or swimming in a public pool.
6. Once you are cured of an STD, you can get it again.
7. The AIDS virus must pass from one person's body fluids into another's bloodstream for the second person to get the disease.
8. Most AIDS patients in the United States are homosexual or bisexual men.
9. AIDS can be spread by heterosexual contact between men and women.
10. AIDS can be spread through contaminated needles.
11. An unborn fetus can be infected with syphilis while in the uterus.
12. An unborn fetus cannot contract gonorrhea while in the uterus, but may contract the disease when an infected mother's "bag of water" breaks before delivery.
13. An estimated 40% of males and 80% of females infected with gonorrhea have no visible symptoms, but they may pass the disease to a sexual partner.
14. If a woman receives no treatment or insufficient treatment for gonorrhea, sterility may occur.
15. A person can still be infected with syphilis even though the syphilic chancre goes away.
16. Syphilis can remain dormant for years and then cause heart or brain damage, or even cause death.
17. Mental illness, blindness, paralysis, and heart disease are all symptoms of syphilis.
18. If a woman has an active case of genital herpes, she may infect her baby during delivery.
19. The doctor of a woman with genital herpes may suggest a cesarean delivery so the baby does not have to pass through the infected birth canal.
20. Women who are infected with genital warts or herpes may develop cancer of the cervix.
21. Condoms help protect against many STDs, but are not 100% effective.
22. If two people are free from STDs and have no other sexual partners, they will likely never have any STDs.
23. Chlamydia can cause pelvic inflammatory disease (PID) which can lead to infertility.
24. Genital warts are caused by a virus and spread by sexual contact.
25. An individual who has been exposed to genital warts may not notice any symptoms for 6–8 months.
26. Some STD strains are becoming resistant to present medications that are available.
27. Many STDs have latent stages where no visible symptoms are present.
28. A baby born to a mother with active genital herpes may not survive, or may be physically or mentally damaged.

Source: www.uen.org/Lessonplan/downloadFile.cgi?file=4357-6-10096-Fact_or_Fiction.pdf&filename=Fact_or_Fiction.pdf

Lesson 4

The Art of Argument

Research and Debate

To the Teacher

During this final phase of the unit, students will be assigned to roles either for or against the mandatory vaccination of 11–17 year olds. They will collaboratively research their sides of the debate, which will be held during the next class period.

Objectives

- Students will review the information they have learned about the human immune system, vaccines, STDs, and cancer.
- Students will produce valid arguments supported by scientific evidence.
- Students will learn about listening to each other's points of view in a critical manner.

Materials

Computers with internet access

Procedure

1. Inform students that they are randomly being assigned to teams for the mandatory vaccine debate. Reassure them that they will have a chance to voice their actual opinions after the debate.
2. Instruct the students of common sides to get together and subdivide the debate into the following smaller groups:
    a. *Opening Statement Group* will make introductory claims present evidence.

b. *Cross Examination Group* will ask clarifying questions of the opposing group (this group must try to anticipate the claims and evidence of the other team and must listen *very carefully* during the debate in order to develop additional questions).

c. *Rebuttal Group* will present rebuttals to the opposing team (this group also must try to anticipate the claims and evidence of the other team and must also listen VERY CAREFULLY during the debate in order to develop additional rebuttals).

d. *Closing Statement Group* will summarize the points and try to make a final, persuasive argument for the "judge."

3. The students should have an opportunity to reconvene with their entire team to review the debate roles.
4. On the day of the debate, allow students 5 minutes to review their roles and notes.
5. Each group will have 3 minutes (for a combined time of 24 minutes for both teams) to make their arguments or pose their questions.
6. At the completion of the debate, distribute the grading rubric to students and ask them to assess the various roles.
7. Compare student assessments with your assessment.

Closure

Ask, "What were the most compelling arguments? Why?" "Did the opposing team make any arguments or present evidence that you did not anticipate?" "What, if anything, surprised you during this process?"

Remind students that they will have a chance to discuss their actual opinions on the issue during the next class.

Assessment

Students will be assessed on the debate rubric (see Debate Assessment Sheet, p 216).

Debate Assessment Sheet

	Affirmative (Vaccine should be mandatory.)	**Negative** (Vaccine should not be mandatory.)
Opening Statement Strength of claims and evidence		
Clarity		
Cross-Examination Strength of questions		
Clarity		
Rebuttal Strength of arguments and evidence		
Clarity		
Closing Statement Strength of arguments and evidence		
Clarity		
Total Score		

Scoring: Scale of 1 (low) to 5 (high)
Maximum Score: 40 points per team

5

Lesson 5

Rethinking Positions and Relating the Gardasil Debate to the Nature of Science (NOS)

To the Teacher

During this final lesson, students will have an opportunity to revisit their original position statement from the first lesson in order to form more complete arguments and reassess their positions. They will also discuss possible compromises in regard to the issue. Finally, students will consider the Nature of Science (NOS) aspects of the Gardasil debate, particularly the qualities of scientific evidence, the source(s) of scientific authority, the differences between scientific and non-scientific reasoning, and the impact of science in society. Students will work in groups to create an NOS/Gardasil poster.

Objectives

- Students will evaluate their previous positions and reflect on any changes.
- Students will submit an updated position worksheet.
- Students will work in groups to discuss NOS and create group posters regarding their conceptions of NOS in relation to the Gardasil controversy.

Materials

- Student original position statement handout and a fresh copy of the handout
- Student NOS reflection handout
- Poster paper
- Markers

Procedure

1. Begin by asking the class to review their original position statement on Gardasil and ask whether their positions on the controversy may have changed from the first day of this unit to the current day. Time will be provided to any students who wish to disclose their responses to the class. Ask, "Are there any 'compromises' that could be made to bring the group to an agreement?" (mention that many states have, "opt-out" policies for families who object to the vaccine, perhaps a narrower age requirement, and so on).
2. Distribute a fresh copy of the position statement and provide students with ample time to give their current position as well as the required 10 statements of evidence or reasoning that support their position.
3. Divide the class into small groups. Assign the groups the task of discussing the NOS topics on the handout (attached resource), which relate to the controversy.
4. Each group will then compose a poster on their ideas regarding the NOS questions they discussed.
5. Groups will present their posters and a closing discussion on NOS will take place.

Closure

Ask, "How did you feel about this unit?" "What is the most important thing you learned?"

Assessment

The students will be evaluated by

- the quality of their position statements,
- their responses to class discussion questions, and
- the NOS group discussions and presentations.

NOS Group Discussion/Presentation Rubric

	1 pt.	2 pts.	3 pts.
Sources of Evidence	Group fails to demonstrate understanding of the importance of credibility of sources.	Group conveys understanding of the analysis of credibility assigned to different sources, but fails to accurately identify factors impacting credibility.	Group conveys understanding of the analysis of credibility assigned to different sources of evidence and is able to accurately identify factors impacting credibility.
Diversity of Conclusions	Group fails to recognize different conclusions on the issue.	Group conveys that different conclusions can be reached on an issue, but attributes it to a lack of valid research.	Group conveys understanding that different conclusions can be validly reached on the same issue.
Scientifc vs. Nonscientific Claims	Group fails to distinguish between scientific and nonscientific claims.	Group attempts to distinguish between scientific and nonscientific claims but does so with inaccuracies.	Group accurately distinguishes between scientific and nonscientific claims.

Nature of Science Discussion Guide

Use these questions to guide your discussion and the creation of your poster.

- What does the controversy regarding Gardasil show us about the nature of scientific claims and their relevance to society? Do some scientific claims hold more weight than others in societal decisions? Do some other claims and forms of reasoning hold more weight in societal decisions?
- There are proponents for mandatory Gardasil vaccinations who hold valid scientific arguments and reasoning, and there are opponents who also have valid scientific arguments and reasoning. Is a differing of scientific inferences and interpretations of data a strength of science or is it a weakness?
- What makes a scientific claim different than a non-scientific claim? What are the necessary components of a scientific claim?
- What are valid sources of scientific information? What is required for valid scientific knowledge to become produced? Does valid scientific knowledge come from a singular authority, with a set method?

"Mined" Over Matter

Should rare Earth elements be mined in the United States?

Unit Overview

In this unit, students will investigate the dynamic interplay between environmental, political, and economic factors involved in deciding whether the United States should begin mining rare earth elements (REEs). REEs are of tremendous interest to students, as they are used in everyday technologies such as computers, cell phones, and TVs. Through hands-on, minds-on activities, students will explore the sources, uses, and environmental threats of mining these materials, while examining the controversial question of whether the United States should become self-reliant in supplying these REEs, regardless of environmental consequences.

Key Science Concepts

Earth's Layers, Plate Tectonics, Rock Cycle, Renewable and Nonrenewable Resources, Sustainability, Environmental Quality, Green Technology

Ethical Issues

Environmental Responsibility, Economics and Environmental Trade Offs, Property Rights, Government Regulation of Public Safety

Science Skills

Classifying, Observing, Understanding Cause and Effect, Communicating Results, Interpreting and Representing Data, Forming Arguments From Evidence

Grade Levels

High School Advanced Environmental Science; can be adjusted for General Biology or Environmental Science

Time Needed

This is a seven-lesson, nine-day unit plan, although it can be adapted for a shorter or longer time period depending on the needs of your students.

Lesson Sequence

Lesson 1. Introduction to Rare Earth Elements

Lesson 2. Rock and Roll: Plate Tectonics and the Rock Cycle

Lesson 3. Elements, Rocks, and Minerals

Lesson 4. Cake Mining: Identification and Reclamation of Mineral Resources

Lesson 5. Digging Deeper: Exploring the Controversy

Lesson 6. The Decision-Making Process

Lesson 7. Debate: Should the United States Mine Rare Earth Elements?

Background on the Issue

Rare Earth Elements (REEs) are used in a wide range of technology that is vital to the U.S. economy, defense, and future in green technology. They are used in many of the new technologies that today's society has become dependent on, and may be vitally important in the development of future technologies. REEs are crucial components in such commonplace items as plasma screen TV's, cell phones, computers, and speakers; as well as in the medical industry and in the military. REEs are also important for the development of green technologies, as they are used in hybrid cars, wind turbines, and solar panels. Currently, China produces more than 95% of the world's supply of REEs. As such, they have market power. Recently, China began limiting its exports of REEs, forcing manufacturers of products that require these materials to slow production. It is believed that China is looking to begin manufacturing these products themselves. If this happens, China would have a monopoly on these products, as no one else would be able to manufacture them without access to refined REEs and consequently, China could set the price of these items. Because of this, many countries are looking into the feasibility of

mining REEs within their borders so that they will not be dependent on China for supply. The problem of mining REEs, however, is that the mining and refining process causes severe environmental impacts. Countries such as the United States must weigh the benefits of having available and relatively cheap technology—with its attendant benefits to the country's economy, defense, energy independence, and the well-being of its citizens—against the environmental impacts of mining and the resulting air and water pollution and habitat destruction.

Connecting to *NGSS*

HS-ESS1-5. Evaluate evidence of the past and current movements of continental and oceanic crust and the theory of plate tectonics to explain the ages of crustal rocks.

HS-ESS3-1. Construct an explanation based on evidence for how the availability of natural resources, occurrence of natural hazards, and changes in climate have influenced human activity.

HS-ESS3-2. Evaluate competing design solutions for developing, managing, and utilizing energy and mineral resources based on cost-benefit ratios.

HS-ESS3-4. Evaluate or refine a technological solution that reduces impacts of human activities on natural systems.

Accommodations for Students With Disabilities

Visual Impairments: Be sure to verbally describe photos and diagrams that are projected onto the screen; since this unit involves extensive research from print and online sources, provide large-print handouts for low vision students, use closed-circuit television for projection, or allow students to do all tasks on large computer monitors with screen magnification or enhancement software; allow visually impaired students to participate fully in the rock cycle simulation, with assistance during the heating phase, as the students can feel the differences in the resulting crayon rocks; during the cake mining, encourage visually impaired students to squeeze the cake "coring" out of the straw to feel for "minerals."

Hearing Impairments: Preview necessary vocabulary (i.e., rocks, minerals, plate tectonics, and so on) with student and interpreter (if one is used); it is essential that interpreters understand the content so that they can accurately convey the information. In addition, the interpreter may need time to review the signs for specific scientific vocabulary; allow for "wait time" after asking questions so that

hearing-impaired students can process information; give instructions in written and verbal formats; be sure to face the student when you are speaking; remind teammates and classmates to face student when speaking; during debates or class discussions, have student speakers give a signal (such as a slight hand wave) to alert student to the source of the sound.

Learning Disabilities: Provide graphic organizers to help students keep track of the facts they learn and the arguments those facts support; provide individual Periodic Tables of Elements that students can write on and identify the REEs; point out specific data on bar graphs and charts to model data analysis for students; use "wait time" for all students; provide students with highlighters, sticky notes, or other aids for organizing notes and readings; display a class T-chart on which students can input "pros" and "cons" of REE mining as they progress through the unit.

Motor/Orthopedic Impairment: Provide large materials (i.e., large rock/mineral samples, chunky crayons for rock cycle activity, large streak plates) for students with low grip strength; Allow student to utilize alternative note-taking devices including voice-to-text software if writing is not feasible; large plastic test tubes (rather than straws) can be used for the cake mining activity; if a student uses a wheelchair, have all students present their findings from a seated position.

Emotional Disabilities: This unit involves several group activities. Remind all students that a large part of the process in working through meaningful socioscientific issues is practicing cooperative skills such as consensus-building, collaboration, compromise, and communication; allow students "time out" if discussions become too heated or behavior becomes inappropriate; allow students to express frustrations or challenges in appropriate ways, such as journaling, drawing, one-on-one conversations, or online discussion boards with you.

Resources for Teachers

- U.S. Department of Energy's Ames Laboratory (*www.ameslab.gov/rare-earth-metals*)
- U.S. Geological Survey (*http://pubs.usgs.gov/fs/2002/fs087-02*)
- Geology.com website (*http://geology.com/rare-earth-elements*)

Note: This unit is based on materials submitted by Crystal Nance

Lesson 1

Introduction to Rare Earth Elements

To the Teacher

In this lesson, students are introduced to Rare Earth Elements (REEs), their products and sources, and the dilemma of the world's current reliance on China for these critical materials. Students begin to examine their opinions on the question of whether REEs should be mined in the United States and conduct a personal mineral consumption inventory.

Topic: Rare Earth Elements
Go to: *www.scilinks.org*
Code: SSI009

Objectives

Students should be able to do the following:

1. Access and explain prior knowledge about elements, minerals, and mining and their connection to products used by society.
2. Demonstrate an understanding of the connections between science, technology, and society.
3. Connect science content learned in the classroom with larger community and societal issues.

Time Needed

One 50-minute class period

Materials

- LCD or overhead projector (or other way to project information)
- Whiteboard to write down student ideas
- Pictures of technology that uses REEs (suggestions: iPod/MP3 player, plasma TV, laptop computer, hybrid car, CFL bulb, cell phone, wind turbine, fiber optics, MRI machine)

- Table of REEs and their uses (Figures 6.1 and 6.2)
- Periodic Chart (on wall)
- Personal Mineral Consumption worksheet

Procedure

1. Project images of the products which require the use of REEs in their manufacture.
2. Ask the students what these products have in common. Accept and write down all answers on the whiteboard.
3. Tell the students that in the very near future, these products may not be as widely available as they are now and they will likely be more expensive. So, when the cell phone or TV or computer they have now dies, it might be too expensive to replace. Ask students to imagine their life without these items—no Facebook, no Twitter, no texting, and so on.
4. Allow students to react and express their feelings and ideas.
5. Explain that each of these products need metals called Rare Earth Elements in order to function. Point out the REEs on the Periodic table. Ask the students what they know about elements to recall prior knowledge.
6. Project a list of REEs and their uses so students become familiar with these 17 elements and the properties that make them unique (i.e., strongly paramagnetic, fluoresces). For example, REEs are used in MRIs because they make strong magnets. Students do not need to memorize this information, just have a working understanding.
7. Have students brainstorm ideas as to where we acquire REEs and why the products they rely on may not be available in the future. Accept and write down all answers on the whiteboard.
8. Explain that REEs are found in mineral ores that can be mined from the earth and that the REEs are extracted and refined to be used in manufacturing these products. Ask students what they know about minerals and mining to recall prior knowledge. Accept and write down all answers on the whiteboard.
9. Ask the students what they think are the environmental consequences of mining. Accept and write down all answers on the whiteboard.

10. Explain that China is currently the producer of over 95% of the world's supply of REEs and that China has decided to limit the export of REEs. This means that manufacturers of the technology they use may not be able to maintain production. Because of this, other countries, including the United States, are looking into the feasibility of mining REEs within their own borders so that they do not have to rely on China for supply. Ask students what they think about this idea. Accept and write down all ideas and opinions on the whiteboard.
11. Explain that in order for the students to have an informed opinion and to come to an informed decision, they first have to understand the science involved. Explain that over the next few days, the students are going to learn about how minerals are formed and distributed throughout the Earth, how geologists go about finding mineral resources in sufficient quantities to mine, different mining and refining techniques, and the environmental impacts of mining. They will then use this knowledge to decide if the United States should mine REEs.
12. Record for yourself a copy of all of the students' ideas and prior conceptions that were written on the whiteboard. Use this to inform lesson planning over the following days. Also use to compare student knowledge at the end of unit to the start of the unit to assess whether students learned what you intended and were able to apply that knowledge to informed decision making. Have students' ideas and opinions changed?

Closure

Consider the following cartoon published by the Minerals Education Corporation. What message do you think it is trying to convey? Are you surprised by the data? How could you confirm the statistics included?

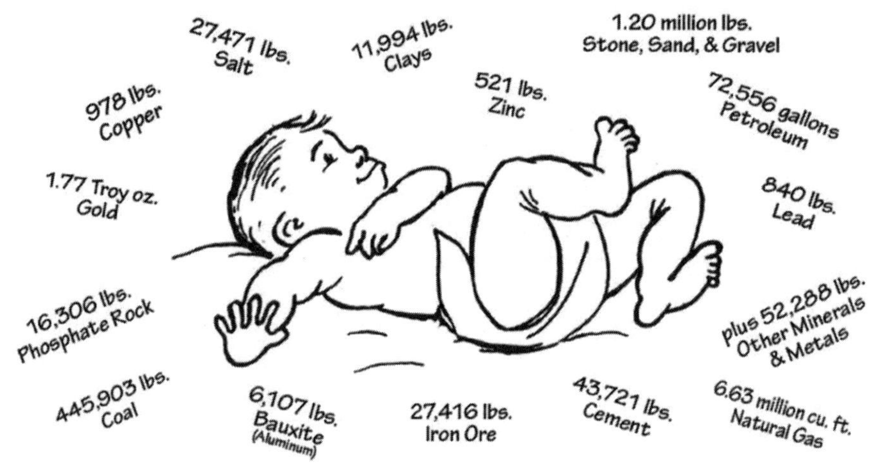

Assessment

Students will be monitored and assessed based on their participation in the class discussion. They will also turn in their Mineral Consumption Worksheet for assessment.

FIGURE 6.1.

The 17 Rare Earth Elements and Their Uses

United States Usage (2008 data)	
Metallurgy & alloys	29%
Electronics	18%
Chemical Catalysts	14%
Phosphors for monitors, television, lighting	12%
Catalytic converters	9%
Glass polishing	6%
Permanent magnets	5%
Petroleum refining	4%
Other	3%

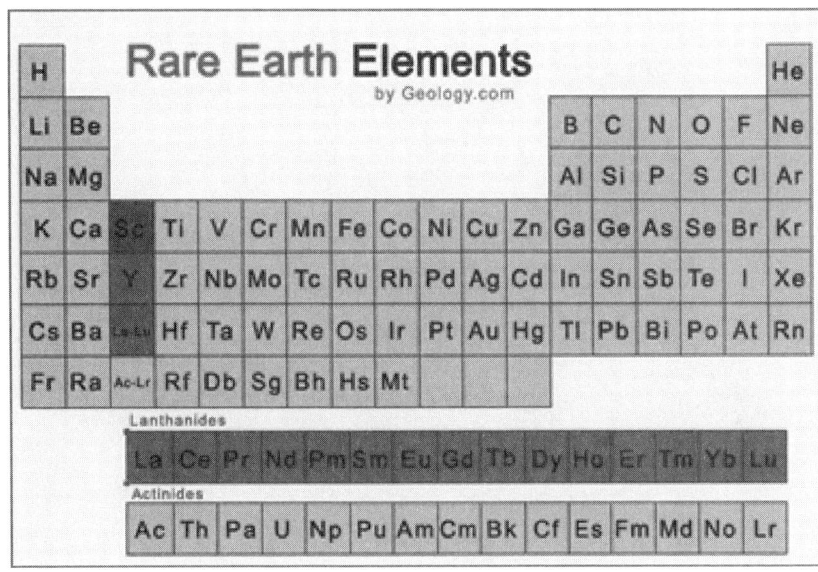

FIGURE 6.2.

Mine Production and Reserves of REEs (2009 Data)

Country	Production (Metric Ton)	Reserves (Metric Ton)
United States	insignificant	13,000,000
Australia	insignificant	5,400,000
Brazil	650	48,000
China	120,000	36,000,000
Commonwealth of Independent States	not available	19,000,000
India	2,700	3,100,000
Malaysia	380	30,000
Other countries	not available	22,000,000
World total (rounded)	124,000	99,000,000

(SOURCE: WWW.GEOLOGY.COM)

Personal Mineral Consumption Worksheet

Objective

Students will calculate the total amount of selected minerals they consume in a lifetime. The exercise will look at mineral production from the "demand side"—what we consume on a yearly basis. The exercise will also help students learn to critically think about how their lives may be affected if the supply (availability) of a mineral resource changes.

Topic: Mining Minerals
Go to: *www.scilinks.org*
Code: SSI010

Instructions

Find the current cost for the commodities listed on worksheet 1 and complete worksheets 1 and 2. Mineral and economic information can be obtained from the following sources:

- Business section of most major newspapers, such as the *New York Times, Wall Street Journal, Financial Times*
- Mineral and economic information from the USGS website
- Doing an online search for a specific commodity

Using Worksheet 1

Step 1: Our society is based on mineral resources. Column A lists eight minerals or commodities that we use on a daily basis, whether metals for machinery, mineral fertilizers for agriculture, or aggregates for construction. Look at the list and name several products that are made from the minerals or commodities listed in column A. Use the internet to identify these products. The values obtained for this exercise are the values listed for each commodity in Worksheet 1 (p. 233). (Note: For the purpose of this exercise, the 1.55 million pounds of stone, sand, and gravel has been combined as one entry—1.55 million pounds of sand/gravel).

Step 2: Column B lists the estimated number of pounds (or metric ton equivalent) of each commodity that a person will use in his/her lifetime. Determine the cost per pound or metric ton for each commodity and enter the number in the appropriate column—column C, cost per pound, or column D, cost per metric ton. Note

that the cost for aluminum, copper, and lead is usually quoted in dollars per pound ($/pound); cost for cement, iron ore, phosphate, salt, and sand/gravel is usually quoted in dollars per metric ton ($/ton).

Step 3: Multiply the price in column B (pounds or metric ton) times either column C (Cost $/pound) or D (Cost $/metric ton) and enter the number in column E. Column E will give you the total value ($) of the commodity you consumed in your lifetime.

Step 4: Next divide column E by 75 (estimated life expectancy) and enter the number in column F. Column F will give you the dollar value for each commodity you use per year.

Step 5: Total the values for column E and column F and enter them at the bottom right of the worksheet.

Using Worksheet 2

Step 1: Write the total value obtained from worksheet 1, column F, in column A of worksheet 2.

Step 2: Write your age in column B.

Step 3: Multiply column A (Total Value used/year) times column B (Your age) and enter the number in column C. Column C represents the dollar amount of the selected commodities you have used since you were born (Total Value used).

Step 4: Subtract your age from 75. Multiply this value with the value in column A and enter it in column D. This will give you the Total Future Value used.

Step 5: Add the values from column C and D together and enter it in column E. Column E represents the Value used in a lifetime. (*Note*: This should be the same value, to the nearest dollar, as the total value for column E on worksheet 1.)

Name: ______________________________ Date: ______________

Worksheet 1

A	B		C	D	E	F
	Amount Used Per American in a Lifetime					
Mineral or Commodity	**Pounds**	**Metric Tons**	**Cost Per Pound**	**Cost Per Metric Ton**	**Value Used in Lifetime**	**Value Used Per Year**
Aluminum	4,864	2.4				
Cement	65,543	32.7				
Copper	1,390	0.69				
Iron Ore	32,810	16.4				
Lead	849	0.42				
Phosphate	21,848	10.92				
Salt	32,061	16.03				
Sand/Gravel	1,550,000	775.00				
				Total:		

List examples of products made from each of the minerals and commodities above:

Name: ______________________________ Date: ____________

Worksheet 2

A	B	C	D	E
Total Value Used PerYear	Your Age	Total Value Used	Total Future Value Used	Value Used in a Lifetime

Lesson 2

Rock and Roll

Plate Tectonics and the Rock Cycle

To the Teacher

During this lesson, students engage with computer simulations and readings to understand plate tectonics and the rock cycle, and model the rock cycle in a hands-on activity.

Topic: Plate Tectonics
Go to: *www.scilinks.org*
Code: SSI011

Objectives

Students should be able to do the following:

1. Draw and label the layers of the Earth.
2. Explain plate tectonics and describe three different types of plate boundaries.
3. Relate rock formation to plate tectonics.
4. Describe three types of rocks and the processes that change one type of rock into another.
5. Investigate and analyze the processes responsible for the rock cycle.
6. Identify forms of energy that drive the rock cycle.
7. Analyze the relationship between the Earth's layers, plate tectonics, and the rock cycle.

Time Needed

One 50-minute class period

Materials

- Computer projector and screen
- Computer stations with internet access
- Copies of the student worksheet
- Different color crayons
- Grater
- Wooden blocks
- Hammers
- Candles or Bunsen burners
- Heavy-duty aluminum foil
- Aluminum pie pan
- Tongs or clothespin

Procedure

1. Ask the students what they remember about the layers of the Earth to draw out their prior knowledge. Draw what they describe on the board with labels. Students will be asked to clarify and revise until the drawing is correct. Go over the layers of the Earth with the students.
2. Show video *An Introduction to Plate Tectonics* from *www.teachersdomain.com*
3. Ask the students the following questions:
    - Is Earth's surface stable and stationary now? Was it ever in the past? Do you think it will be stable in the future?
    - Do you think that the way continents fit together is convincing evidence for the theory of plate tectonics? Why or why not?
    - What was the supercontinent called that once contained nearly all of the continental crust? What do you think the name means?
    - Which ocean is growing in size? Which is shrinking? Explain why this is occurring.

- How does the theory of plate tectonics help us explain natural phenomena such as earthquakes and mountains, which geologists had difficulty accounting for prior to the development of the theory?

4. Divide students into groups and have them work together at computer stations to use the interactives "Mountain Maker, Earth Shaker" and "The Rock Cycle" from *www.pbslearningmedia.org*.

5. As a group, the students will answer the questions on the attached worksheet in their notebooks.

In their groups the students will model the rock cycle by following the directions provided on An Introduction to Plate Tectonics student worksheet.

Assessment

Students will be assessed based on how well they worked together and stayed on task while at the computer stations. The answers to the questions on the worksheet and the homework assignment will be turned in and assessed as well.

An Introduction to Plate Tectonics

After reading the "Background Essay" and exploring the accompanying "Mountain Maker, Earth Shaker" Interactive (*www.pbslearningmedia.org/resource/ess05.sci.ess.earthsys.shake/mountain-maker-earth-shaker*), answer the following questions in your notebook:

1. How do scientists explain why, if new crust is constantly being formed from volcanic materials, Earth's crust stays the same size?
2. Why are continental volcanoes associated with oceanic-continental convergent boundaries?
3. Why are the Appalachian Mountains not as high as the Himalayan Mountains even though they were formed in the same way?
4. Why do scientists predict that Los Angeles will be north of San Francisco in 16 million years?
5. After reading the "Background Essay" and investigating the "Rock Cycle Animation" Interactive at *www.pbslearningmedia.org/resource/ess05.sci.ess.earthsys.rockcycle/rock-cycle-animation*, answer the following questions in your notebook:
6. What is the possible journey of rock material from a lava flow back to magma?
7. How is metamorphic rock formed?
8. How are igneous, sedimentary, and metamorphic rock related?
9. How long do you think the rock cycle has existed?

Modeling the Rock Cycle Activity

1. Remember safety first, so always wear your safety goggles.
2. Lay one piece of heavy-duty aluminum foil of about 20 cm × 10 cm on top of another piece of aluminum foil about the same size.
3. Select crayons to represent igneous rock (we know it is igneous because it lacks streaks, layers, or grains). Using a grater, make enough crayon shavings of different colors to represent sediment. (The grating represents weathering). There should be enough shavings to make a pile 6 cm × 6 cm and 1–2 cm thick.
4. Make a packet of "sediment" within the aluminum foil.
5. Place the packet of aluminum foil between the two wooden boards and press on the top board. Open the packet. What type of rock does this represent? (sedimentary)
6. Rewrap the packet of aluminum foil. Replace the two boards and hammer the top board with more pressure than before. Open the packet. Then place the packet over the heat source using your tongs until the crayon just starts to melt. What type of rock does this represent? (metamorphic—formed by heat and pressure)
7. Place packet over a heat source using your tongs until the colors are completely mixed. Let cool for another few minutes. What type of rock does this represent? (igneous)
8. How could you create sedimentary rock from the igneous rock? (make new shavings from the new "igneous" crayon).

Lesson 3

Elements, Rocks, and Minerals

To the Teacher

In this lesson, students will devise tests to distinguish among rocks and minerals based on their properties, including hardness, streak color, and reactivity with acid.

Objectives

Students should be able to do the following:

1. Compare and contrast rocks, minerals, and elements, and understand that rocks are made of minerals, which are made of one or more elements.
2. Identify and classify different rocks and minerals.

Time Needed

One class period

Materials

- Examples of different rocks of the three types (igneous, sedimentary, metamorphic)
- Examples of many different minerals (the rocks should contain the mineral examples)
- Magnifying lens
- Porcelain streak plates
- Pennies
- Steel files
- Glass plates
- Weak HCL

- Droppers
- Mohs Hardness Table
- Goggles

Procedure

1. Show students the different minerals available. Hold up a mineral and ask students what it is that you are holding. You are looking for "mineral" as the answer. If they say rock, hold up a rock and ask them what it is. How is it different than the first thing you held up? If they say a specific mineral, such as quartz, ask them what quartz is. You want them to eventually say mineral. Once someone gives this answer, ask, "What is a mineral?" "What are minerals made of?" "What have we already learned about minerals?" Once again, the purpose is to draw out the students' prior knowledge.
2. Divide students into groups. Each group will be given 10–12 different examples of minerals, a magnifying lens, a porcelain streak plate, a glass plate, a penny, a steel file, and weak HCL solution and a dropper.
3. Using any or none of the materials available, using any method they devise, challenge the students to come up with a variety of ways to tell the minerals apart. Do not tell them how to use the tools.
4. Have the students keep a data table of their observations and tests and to come up with the best method for classifying the minerals. How is this like how scientists do science?
5. Have students share their "best method" with the rest of the class, explaining why they think it is the best method.
6. Ask the audience members to write down the methods as they are presented and to list one strength and one weakness related to each method.
7. Once all groups have shared, create a master list of the methods with their strengths and weaknesses. Discuss why no one method is optimal.
8. Describe the traditional methods (i.e., Mohs Hardness Scale, streak color, reactivity with HCL) that scientists use to identify and classify minerals and have students use these methods to identify their specimens. (You should know what each mineral is so you can determine if the groups come to the correct answer.)

9. After the groups have had a chance to identify all of the mineral specimens, go over with the whole class the name of each, what elements make up the mineral (thereby highlighting again the fact that minerals are made up of one or more elements), and any other interesting facts about the mineral.
10. Hand out the rock specimens to the groups. Ask the groups to identify the rocks as sedimentary, metamorphic, or igneous. How can they tell?
11. Ask them to use the magnifying lens to study the rocks closely. What do they notice? Ask them to record their observations.
12. Explain to them that rocks are made up of one or more minerals. Challenge them to see if they can tell what minerals their rocks are made of. Which type of rock is it easiest to do this with? Why might this be?

Assessment

As the students are working in their groups, they will be assessed based on how well they work together, and on participation and focus.

Scoring Rubric

	1 pt.	2 pts.	3 pts.
Group Collaboration	Participants rarely demonstrated positive communication and collaboration skills	Participants demonstrated positive communication and collaboration skills most of the time.	Participants consistently demonstrated positive communication and collaboration skills
Group Participation	The group rarely participated in class discussions	The group participated thoughtfully in class discussions but only certain members participated	The group participated thoughtfully in class discussions with all members contributing
Group Productivity	The group frequently got off task and failed to demonstrate some key understandings	The group occasionally got off task but still demonstrated key understandings	The group stayed on task throughout the activity and demonstrated key understandings

Homework

Assign pages 1–6 in the U.S. Geological Survey's "The Life Cycle of a Mineral Deposit Teacher's Guide (available at *http://pubs.usgs.gov/gip/2005/17/gip-17.pdf*) for homework in preparation for the next class.

6

Lesson 4

Cake Mining

Identification and Reclamation of Mineral Sources

To the Teacher

In this lesson, students apply what they have learned about mining to a simulation utilizing cake and various items to represent ores and other earth materials. Students develop mining and reclamation plans, and gain understanding of the economic and environmental impacts of mining.

Objectives

Students should be able to do the following:

1. Describe the different methods geologists use to find and mine mineral resources.
2. Develop and implement a plan to explore and profitably mine an area for minerals.
3. Develop and implement a reclamation plan to restore a mined area to its natural ecology.
4. Evaluate their plans and the plans of others for economic and environmental soundness.

Time Needed

One class period

Materials

- Cake Mining Activity Handout

- Decorated marble cake mining area (see "Teacher Preparation")
- Clear straw for core sampling
- Tool kit to mine minerals (toothpicks, spoons, forceps, and so on)
- Play money

Teacher Preparation

In order to create a "cake mine," prepare a rectangular marble cake according to package directions, but be sure to include candy sprinkles in the dark batter. The candy sprinkles will represent minerals. Frost the cake so that the top of the "land" underneath isn't visible. If you wish, you can make the land more realistic by coloring the frosting green for vegetation, adding blue areas for water and pretzels for trees, and so on.

Topic: Rock Cycle
Go to: *www.scilinks.org*
Code: SSI012

Procedure

1. Review the homework reading by asking the following questions and writing the response on the whiteboard:
   a. What are some ways geologists go about looking for areas that may contain mineral resources to mine?
   b. What are some methods of mining? Describe them. When are they appropriate for use?
   c. What are some of the methods used to refine minerals or extract them from the ore?
2. Describe how REEs are mined and processed. (Open-pit or sometimes open-pit strip). Use the EPA report (*http://nepis.epa.gov/Adobe/PDF/P100EUBC.pdf*) as a reference.
3. Divide the students into groups and give them the Cake Mining Activity handout.
4. Go over the procedure for the activity with the students before they begin.
5. As the groups are working, monitor their progress and review their plans.
6. Once the groups have mined their minerals and reclaimed their piece of land, have the groups present their plans to the rest of the class and describe

whether they felt their plans were successful or not, and what, if anything, they would do differently next time.

7. While the students are listening to the other groups present, they can eat their project.

Assessment

Students will be assessed according to their level of participation in the class discussion. When working on the cake activity, students will be assessed according to how well they work together, time on task, the thoughtfulness of their plans, and their group presentations.

Scoring Rubric

	1 pt.	2 pts.	3 pts.
Group Collaboration	Participants rarely demonstrated positive communication and collaboration skills.	Participants demonstrated positive communication and collaboration skills most of the time.	Participants consistently demonstrated positive communication and collaboration skills.
Group Mining Plans	The group failed to develop or analyze a thoughtful mining plan.	The group developed a thoughtful mining plan but failed to critically analyze its success.	The group developed, followed, and evaluated a thoughtful mining plan.
Group Productivity	The group frequently got off task and failed to demonstrate some key understandings.	The group occasionally got off task but still demonstrated key understandings.	The group stayed on task throughout the activity and demonstrated key understandings.

Cake Mining Activity Handout

You are with a mining company and you have just received rights to mine an area for REEs.

1. The object of mineral mining is to make a profit. You will start with $25 and you must use this money to buy the mining property and the mining tools, and to pay the mining, refining, and reclamation costs. In return you will receive money for the minerals you mine and refine. Costs are as follows:

   a. 2″ × 2″ piece of cake is $3

   b. 3″ × 3″ piece of cake is $4

   c. 4″ × 4″ piece of cake is $5

   d. A piece of clear straw is $1

   e. A toothpick is $1

   f. A forceps is $2

   g. A spoon is $3

   h. Mining cost is $4

   i. Refining cost is $4

   j. Reclamation cost is $6

   (If a piece of equipment breaks, it is no longer useable, and you must buy another if you want to replace it.)

2. Decide what size land and what tools you want and make the purchases. Record your purchases and the money you spent in your notebook. Take your items to your table.
3. Investigate your land. Can you tell from the surface where your mineral ore is located?
4. Develop a plan for exploring for the location of your mineral resources. Your land is made of marble cake. The dark areas are the mineral ore and the light areas do not contain mineable minerals. (Hint: You could use the straw tool to take core samples.)
5. After you have determined where your mineral resources are located, develop a plan for mining your mineral ore. You must write out your plan and get it approved before you begin. You may only use the tools you have purchased to mine your land. You cannot use your hands! Pay your mining fee of $4.
6. Before mining, you must also develop a reclamation plan. This plan must include how your company will restore the land to the way it was prior to mining. You must write out your plan and

get it approved before you begin. You may only use the tools you have purchased to restore your land. You cannot use your hands! Pay your reclamation fee of $6.

7. Once both your mining plan and reclamation plan has been approved, you may begin.
8. After you have mined your ore, you also need to develop a way to extract the mineral containing REEs (candy sprinkles) from the ore (dark marble cake). You may only use the tools you have purchased to refine your minerals. You cannot use your hands! Pay your refining fee of $4.
9. Sell your minerals. Each mineral (candy sprinkle) will earn you $5. Write in your notebook how many you recovered and how much money you earned from mining. Did your mine make a profit?
10. When finished, be prepared to explain the procedures you used to mine and refine your minerals containing REEs, and to restore your land. Did you succeed at gathering all of the minerals in your land? Were you able to make a profit? Was your reclamation plan successful? Can the area of land that you mined ever be returned to how it was originally? What would you do different if you were to do this activity again?

Lesson 5

Digging Deeper

Exploring the Controversy

To the Teacher

In this lesson, students will perform research on the issue from primary sources and share the information with classmates using a jigsaw cooperative learning strategy.

Objectives

Students should be able to do the following:

1. Extract and compile pertinent information from a variety of sources.
2. Clearly and accurately share what they learned with others.

Time Needed

Two class periods

Materials

Readings (See the resource list at the end of this lesson.)

Procedure

1. This lesson is a Jigsaw activity.
2. Divide the students into expert groups. Give each expert group one or two readings from the resource list on which to become experts. Instruct the groups to carefully extract as much information as they can from the resource and to record that information. They will need to know it well enough to teach others.

3. Provide as much time as the groups need to become experts on the resources and to compile a list of "facts." The groups should also note such things as the types of information presented (scientific, economic, and so on), the source of information, and what bias may be present.
4. Once the expert groups are finished, divide the students into new groups where each group has one expert from each of the expert groups.
5. Each expert will share what he or she learned with the other students in the new group.
6. Instruct the new groups to compile what they learned and prepare a presentation. Let the students know that you will be assessing the presentations for completeness and accuracy, so they need to be conscientious with their work.
7. After the groups have had sufficient time, have each group present what they learned. Instruct the students that if a group presents a point that they missed, they should be sure to take note of it.
8. Each student should have a copy of all of the information that has been compiled.
9. The goal of this lesson is for each student to be fully versed in all sides of the socioscientific issue of whether REEs should be mined in the United States.

Assessment

Students will be assessed according to their participation and how well they worked together. The group presentations will also be assessed for completeness and accuracy.

Expert Group Scoring Rubric

1 pt.	2 pts.	3 pts.	4 pts.	5 pts.
Expert groups conduct limited, inaccurate research.	Expert groups conduct thorough research that contained inaccuracies.	Expert groups conduct accurate research that is presented in a clear manner, yet some important information was missing.	Expert groups conduct accurate and thorough research; however, the findings are presented in an unclear or unpersuasive manner.	Expert groups conduct accurate, thorough research and present their findings in a clear and compelling manner.

Resources

Avalon Rare Metals, Inc. Rare Earths 101. *http://avalonraremetals.com*

Center for Strategic and International Studies. 2010. Rare earth elements: A wrench in the supply chain? October. *www.csis.org/isp/diig*

Comisso, K. 2010. U.S. reserves of rare earth elements assessed for the first time. *New Scientist* November 19. *www.newscientist.com/article/dn19753-us-reserves-of-rare-earth-elements-assessed-for-first-time.html*

Environmental Protection Agency (EPA). 2011. Investigating rare earth element mine development in EPA region 8 and potential environmental impacts. *www.epa.gov/region8/mining/ReportOnRareEarthElements.pdf*

Hurst, C. 2010. China's rare earth elements industry: What can the west learn? *Institute for the Analysis of Global Security.* March. *www.iags.org*

Staufer, P. H., and I. I. Hendley, ed. 2002 Rare earth elements: Critical resources for high technology. USGS fact sheet. *http://geopubs.wr.usgs.gov/fact-sheet/fs087-02*

Tabuchi, H. *New York Times.* 2010. Japan Recycles Minerals From Used Electronics. Oct 4. *www.nytimes.com/2010/10/05/business/global/05recycle.html*

Tyler, P. *New York Times.* 2010. Rush on for "rare earths" as U.S. firms seek to counter Chinese monopoly. July 23. *www.nytimes.com/gwire/2010/07/23/23greenwire-rush-on-for-rare-earths-as-us-firms-seek-to-co-58814.html*

United States Government Accountability Office. 2010 Rare earth materials in the defense supply chain. GAO report # GAO-10-617R. *www.gao.gov*

Lesson 6

The Decision-Making Process

To the Teacher

During this lesson, students will use a compensatory decision-making process to evaluate the pros and cons of REE mining. This process will aid students as they prepare for a debate on the issue of whether the United States should begin mining rare earth minerals in order to become self-reliant in supplying this resource.

Objectives

Students will be able to do the following:

1. Identify and evaluate all sides of an issue based on a variety of sources and types of information, especially scientific.
2. Identify criteria important in arriving at an informed decision.
3. Justify decisions based on sound logic and reasoning.

Time Needed

One 50-minute class period

Materials

- Computer with internet access
- Compensatory Decision-Making Worksheet

Procedure

1. Randomly assign students to a position by picking it out of a hat or box. The positions are as follows:
    a. Green tech small business owners (need REEs to manufacture products and develop new technologies)

   b. Community citizens (live near proposed mine)

   c. Mining companies (contracted to being REE exploration and mining)

   d. Environmental groups

   e. U.S. military (use REEs in equipment; national security concerns)

   f. Politicians (concerned with foreign policy and relations regarding China, in particular)

2. Remind the groups that the question is "Should REEs be mined in the United States?"

3. Each position group will meet to discuss what its opinion is regarding the issue and to discuss differences.

4. Go to each group, one at a time, and ask students where they are at in their decision-making process. Hopefully, they will see that there are "good" and "bad" sides to whatever decision they make, which makes the decision-making process difficult.

5. Explain the compensatory decision-making process to the group where the group will try to balance the pros and cons to see which decision comes out better (see below). The group will be instructed to use this decision-making framework to come to a conclusion regarding where they will stand regarding the issue.

   a. The group will identify criteria that are important factors that affect its decision and make a table with columns under "yes" or "no" and rows labeled with its criteria.

   b. The group will then allocate 10 points between the "yes" and "no" columns based on how well it fulfills the criterion.

   c. The groups will then add up the numbers under each column of "yes" or "no," and the one with the highest number of points is the decision based on this method.

   d. If one criterion is deemed more important than any other, or one particular decision had no negative outcomes, the group may decide to base their decision only on that one criterion or go with that particular decision (non-compensatory strategy). Let the groups know that this is fine too, but no matter how they arrive at their decision, they must be able to explain and justify it.

   e. The students also have the option of rejecting the decision the process led them to, but again, they must be able to justify their reasons for doing so.

6. Allow the groups as much time as they need to come to a decision. Once they have done so, ask them to call you over so that they can explain their decision-making process and justify their final decision to you. Ask them if their decision is different now than what they were thinking originally. How or how not? If it is different, what made them change their mind? Do they think this process was valuable? Do they think they might apply the decision-making framework to other decisions in the future?

Closure

Ask students to think of another decision they are facing in their lives, small or large, and to apply the decision-making process they learned in class to it. Do they think this is a valuable strategy? Why or why not?

Assessment

Students will be assessed based on how well they work together and their participation in the group decision-making activity. Each group will also be evaluated based on how well they applied the decision-making process to the SSI question.

Scoring Rubric

	1 pt.	2 pts.	3 pts.
Group Collaboration	Participants rarely demonstrated positive communication and collaboration skills.	Participants demonstrated positive communication and collaboration skills most of the time.	Participants consistently demonstrated positive communication and collaboration skills.
Group Application of Compensatory Decision Making	The group failed to reach a decision utilizing compensatory decision-making strategies.	The group group reached a decision utilizing compensatory decision-making strategies but did not analyze the process.	The group reached a decision utilizing compensatory decision-making strategies and thoughtfully analyzed the process.
Group Productivity	The group frequently got off task and only some members contributed to the decision-making process.	The group occasionally got off task but all members contributed to the decision-making process.	The group stayed on task throughout the activity and all members participated in the decision-making process.

Name: ______________________________ Date: ____________

Compensatory Decision-Making Worksheet

Should REEs Be Mined in the United States?

Criteria	Yes	No	Total Points
			10
			10
			10
			10
			10
			10

Total Points ________

1. In your team, determine the list of criteria that will inform your decision on the issue and enter them in the "Criteria" column.
2. Allocate a total of 10 points between the "Yes" and "No" columns to indicate the level of agreement or disagreement for each criteria on the question of whether REE's should be mined in the U.S. (For example, if your criterion is "Increased Job Opportunities," determine how many points out of 10 this criterion supports a "Yes" or "No" vote).
3. When completed, add up the "Yes" and "No" columns to determine your team's vote on the issue. The highest score represents the winning decision.
4. Note that if your team believes that one criterion outweighs all the rest (a "non-negotiable"), you can vote in a non-compensatory fashion based on that one criterion. You can also decide to reject the outcome the compensatory process led to. But you must be able to explain and justify your decisions.

Lesson 7

Debate: Should the United States Mine Rare Earth Elements?

To the Teacher

During this lesson, students will engage in a semistructured debate on the question of mining rare earth elements.

Objectives

Students will be able to do the following:

1. Present a clear and well-reasoned argument based on logic and science.
2. Conduct themselves in a respectful manner during argumentation.
3. Describe how being scientifically literate and having a deep conceptual understanding of scientific concepts helps them to make more informed decisions.
4. Describe to what extent science informs, and is informed by, societal concerns and needs.
5. Connect science content learned in the classroom with larger community/societal issues.
6. Demonstrate scientific habits of mind, critical thinking, and decision-making skills.

Time Needed

Two class periods (one for debate preparation and one for the debate).

Materials

- Timer for 10-minute opening statements
- Debate Assessment Sheet

Procedure

1. Instruct the position groups that they are going to develop a statement reflecting their decision on the debate question. Remind them that they must anticipate and prepare for whatever points the other groups may bring up and prepare counterarguments. The students will have a full period to prepare. Monitor the groups and provide guidance as needed.
2. The following class period, the students will debate the topic with the goal of coming to consensus on the question "Should REEs be mined in the United States?"
    a. Explain the rules for the debate and especially regarding conduct. Remind students to treat each other with respect and tolerance for differing opinions.
    b. Each group will choose one person to present two-minute opening statements while the other groups listen respectfully.
    c. The floor will then be opened up to anyone who would like to cross-examine or rebut other teams.
    d. Allow teams to make closing statements if they wish.
    e. Monitor the interactions, but do not interject unless absolutely necessary to halt rude behavior or to get the argumentation back on track.
3. When there is only approximately 10–15 minutes left of class, call the debate to a close and ask if they think they were able to come to a consensus. Have a final vote, yes or no.

Closure

Discuss student reactions to the argumentation process. To what extent does this demonstrate the connections between science, technology, society, and the environment? Does science inform society's decision making? Should it? How does understanding science help them to make more informed decisions in their own lives?

Assessment

Instruct the students to write a three to five page reflection paper on what they have learned during this unit, what their thoughts and opinions on the topic are

(with examples and justifications), and how what they learned in this unit may or may not affect what decisions they make in the future.

A debate assessment sheet will be kept to keep track of who participates in the debate and the quality of their arguments. The debate and the reflection paper will be used as summative assessments.

Name: ______________________ Date: ____________

Debate Assessment Sheet

	Green Technology Small Business Owners	Community Citizens	Mining Companies	Environmental Groups	U.S. Military	Politicians
Opening Statement Strength of claims and evidence						
Clarity						
Cross Examination Strength of questions						
Clarity						
Rebuttal Strength of arguments and evidence						
Clarity						
Closing Statement Strength of arguments and evidence						
Clarity						

Total Score ________

Scoring: Scale of 1 (low) to 5 (high)
Maximum Score: 40 points per team

Unit 7

"Pharma's" Market

Should prescription drugs be advertised directly to consumers?

Unit Overview

Topic: Prescription Drugs
Go to: *www.scilinks.org*
Code: SSI013

In this unit, students will be introduced to the controversy surrounding "direct to consumer" (DTC) advertising of pharmaceuticals. While learning key chemistry concepts related to molecular structures and chemical reactions, students will examine current regulations and processes involved in the research, development, manufacture and distribution of pharmaceutical drugs in the United States. This approach has the benefit of encouraging students to develop argumentation and discourse skills while investigating complex scientific, moral, and political questions.

Key Science Concepts

Solubility, Polarity, Synthesis Reactions, Molecular Structure, Intermolecular Forces

Ethical Issues

Patient Rights, Freedom of Speech, Cost-Benefit Analysis in Health Care, Government Regulation, Paternalism, Free Market Systems

Science Skills

Data Collection and Analysis, Inferring, Developing Arguments From Evidence, Understanding Cause and Effect

Grade Level

High School

Time Needed

This unit is designed for approximately 10 50-minute class periods.

Lesson Sequence

Lesson 1. We Are Family: Prevalence of Prescription Drug Use in the United States

Lesson 2. What's in the Bottle? Research on Pharmaceutical Composition

Lesson 3. How Do Drugs Work? Molecular Models and Drug-Target Interactions

Lesson 4. How Are Drugs Made? Creating a Synthetic Drug

Lesson 5. Vioxx: A Case Study

Lesson 6. Congressional Subcommittee Hearing: Should Prescription Drugs Be Banned From TV?

Background on the Issue

Prescription drug use is on the rise in the United States, according to a study by the Centers for Disease Control (CDC 2010), which found that 48.5% of Americans had used at least one prescription medication within the last month before the survey, and many took five or more prescriptions. Pharmaceuticals are big business, and studies have suggested that pharmaceutical companies spend more on advertising to physicians and to the public than on basic research (York University 2008). Only the United States and New Zealand allow pharmaceutical companies to advertise directly to consumers. In the United States, direct to consumer (DTC) advertising was unregulated until 1962, when it became the purview of the U.S. Food and Drug Administration, which ensures that the advertisements are not misleading. DTC advertising raises the costs of pharmaceuticals tremendously over pharmaceuticals that are not advertised. Proponents of the DTC advertising argue that they empower consumers to have personal choice, to work with their doctors as a team, and are educated about potential illnesses they might have. They also argue that they have freedom of speech under the First Amendment of the Constitution to advertise as long as it is not prohibited (as is tobacco), and that pharmaceutical companies have a right to make profits, particularly given the large investment in

research, and that their success helps the economy. Those against DTC advertising cite the high costs that are passed on to the consumers and the fact that consumers request medications from their doctors based on the ads and not based on medical knowledge. They also argue that the ads are misleading, not adequately policed by the FDA, and cause consumers to think that they have various ailments that need medication when in fact they do not. Moreover, detractors cite that certain drugs (Vioxx, for example), have been advertised to the public before adequate testing has occurred, causing much more harm to the public than if the product's advertising had been delayed, since side effects sometimes do not become apparent during the clinical trials and are more pronounced when used by the public. Therefore, the question of whether DTC advertising should be allowed is one fraught with a great deal of emotional, economic, and political will on both sides.

This unit includes several instructional methods including primary source analysis, hands-on experimentation, Jigsaw readings, computer simulations, and a mock congressional hearing to illuminate the rich scientific content and ethical dilemmas entwined in this controversial question.

References

Centers for Disease Control (CDC). 2010. Prescription drug use continues to increase: U.S. prescription drug data for 2007–2008. NCHS Data Brief. September 2010. *www.cdc.gov/nchs/data/databriefs/db42.pdf*

ProCon.org. *http://prescriptiondrugs.procon.org*

York University. 2008. Big pharma spends more on advertising than research and development, study finds. *ScienceDaily*. January 7. *www.sciencedaily.com/releases/2008/01/080105140107.htm*

Connecting to *NGSS*

HS-PS1-1. Use the periodic table as a model to predict the relative properties of elements based on the patterns of electrons in the outermost energy level of atoms.

HS-PS1-2. Construct and revise an explanation for the outcome of a simple chemical reaction based on the outermost electron states of atoms, trends in the periodic table, and knowledge of the patterns of chemical properties.

HS-PS1-3. Plan and conduct an investigation to gather evidence to compare the structure of substances at the bulk scale to infer the strength of electrical forces between particles.

Accommodations for Students With Disabilities

Visual Impairments: Create a tactile version of the CDC graph using puff paints: letters and lines can be raised by drawing with glue and letting it dry; create braille labels for the pharmaceutical bottles or in the alternative, use a pharmacy website and screen reader software to have student read the label online; have student measure powders (such as the salicylic acid) over aluminum foil so that spillage is audible and use glue dots to make graduated cylinders tactile for liquids; provide or have class create a wide range of tactile molecular models that remain available for students with visual impairments to examine.

Hearing Impairments: Remember that students with hearing impairments who are either lip reading or signing cannot listen to instructions and write or perform an activity at the same time, as you must therefore try to teach sequentially; give instructions first (orally and in text) and then allow students to work. This will be particularly important during the model building and "drug development" lab lessons.

Motor/Orthopedic Disabilities: Provide larger empty bottles/boxes during the "What's in the Bottle?" activity for students with fine-motor skills challenges; similarly, provide large materials such as large clay balls or regular-size marshmallows to create molecular models. These models can also be made on felt boards, magnetic boards, or using computer simulation; make sure that the laboratory workspace is at the appropriate height for a student who uses a wheelchair.

Learning Disabilities: Provide several tangible examples of decimal/percentage/fraction equivalents (such as, "How many students in the class are wearing red?" "How can we express that as a fraction?" "A decimal?" "A percent?") so that students fully comprehend the computation of prevalence of pharmaceuticals during the "We Are Family" activity; pre-teach vocabulary including *polarity, synthesis,* and *bonding* and have students create a graphic organizer (i.e., a T-chart, concept map) to reinforce.

Emotional Disabilities: Familiarize yourself for possible triggers of student behavior such as loud music or sounds and adjust the environment accordingly. If loud music is a trigger, don't play the "We Are Family" song in the first activity; when students are creating molecular models, be sure to assign roles or materials so that each student is included and disputes over materials or jobs are minimized; encourage collaboration during laboratory and subcommittee meetings by identifying a social skill, such as communication, cooperation, or coming to consensus, and rewarding points for positive behaviors.

Resources for Teachers

- National Center for Biotechnology Information on Direct to Consumer Pharmaceutical Advertising (*www.ncbi.nlm.nih.gov/pmc/articles/PMC3278148*)
- Oxford University Press – Drugs and Drug Targets: An Overview *http://fdslive.oup.com/www.oup.com/academic/pdf/13/9780199697397_chapter1.pdf*
- U.S. Food and Drug Association: Oversight of Direct to Consumer Pharmaceutical Advertising (*www.fda.gov/ForConsumers/ConsumerUpdates/ucm107170.htm*)

Note: This unit was developed based on a project submitted by Jessica Croghan-Ingraham with contributions from Kory Bennett and Daniel Majchrzak.

Lesson 1

We Are Family

Prevalence of Prescription Drug Use in the United States

To the Teacher

In this lesson, students are introduced to some of the social and ethical aspects of the current state of the pharmaceutical industry as well as the prevalence of the usage of pharmaceuticals in the United States to show the students that this is relevant to them and their families, now and in the future.

Objectives

Students will be able to analyze statistical data on the prevalence of pharmaceutical use in the United States and calculate a "pharma factor" for their families. They will be able to identify some of the salient issues surrounding pharmaceutical advertising.

Time Needed

One 50-minute class period

Materials

- "We Are Family" worksheet
- projector to display CDC graphic
- recording of "We Are Family" by Sister Sledge (optional)

Procedure

1. Have the song "We Are Family" by Sister Sledge playing as students enter the classroom.
2. Instruct students to complete the "We Are Family" worksheet that surveys the number of their immediate and extended family members by age group.
3. Once students have their totals, show them the CDC graphic of the prescription drug usage by age group. Total the percentage of people in each group who take any number of prescription drugs. Make sure to help them conceptualize these numbers. "1 in 5" may be easier for them to understand than 20%. Show them how to use the decimal form of the percentage to create a "Pharma factor."

FIGURE 7.1.

Percentage of Prescription Drugs Used in Past Month, by Age, United States, 2007–2008

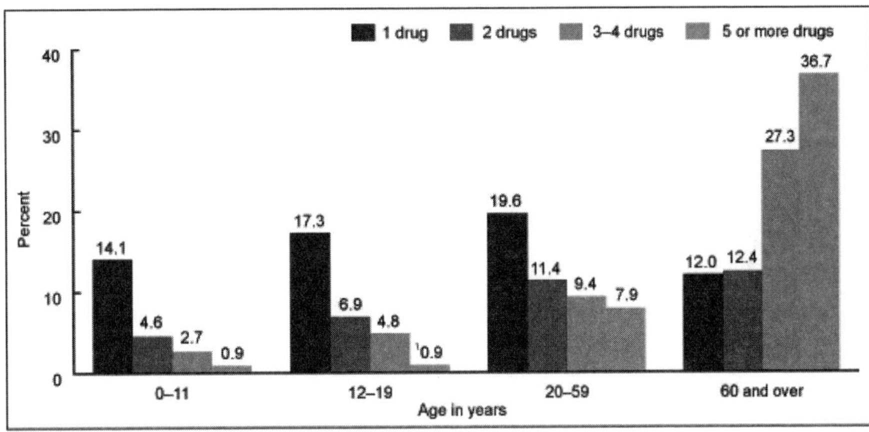

[1]Estimate is unstable; the relative standard error is greater than 30%.
SOURCE: CDC/NCHS, National Health and Nutrition Examination Survey.

Source: www.cdc.gov/nchs/data/databriefs/db42.htm

4. Once students have an understanding of the number of pharmaceutical drugs that are taken in the United States based on age group, moderate a whole-class general discussion about some of the issues related to the pharmaceutical industry. Since the students may have very little concrete knowledge of the regulations that govern pharmaceuticals, the costs to consumers, or the politics involved at this point, this should be an informal discussion that brings up questions that students would like answered in the course of this unit. Here is a possible list of discussion topics:

- In recent years, growth in prescription drug spending has outpaced that of every other category of health expenditures. Drug companies in the United States are allowed to market directly to consumers through ads on TV, radio, and magazines. The United States is one of the few countries that allows this practice. There are some who claim that this has lead to more patients seeking to take prescription medications for a variety of issues. If prescription drugs have to be prescribed by a physician for a medical reason, what do you think the function is of these "direct-to-consumer" ad campaigns? What are the benefits and risks?
- The pharmaceutical industry drug representatives spend a lot of time and money to market their drugs to doctors. Many think that conflicts of interest like this are a dangerous combination when healthcare is involved.
- "Among the most prevalent conflicts of interest are those arising from physicians' interactions with drug company sales representatives, or "detailers." Pharmaceutical companies employ about 90,000 detailers and spend over $7 billion annually to market their products to physicians, averaging $15,000 per year per physician. Prescribing decisions can become conflicted by free gifts, meals, travel, and other benefits. Because physician–detailer interactions bias medical decision-making, undermines public trust, and increase healthcare costs, the medical profession is now under unprecedented pressure to recognize, disclose to the general public, and deal with conflicts of interest." (2007 *Journal of General Internal Medicine* Feb. 22 (2): 184–190.)
- Patents for pharmaceuticals last about 20 years. After that time, generic drug or "copy-cat" drug companies can manufacture the drug for a reduced cost to consumers. When a pharmaceutical company discovers a "blockbuster" drug, they make billions of dollars during the time that their patent is in effect. What are the benefits and costs of having the system set up this way?
- "Patent protections encourage research and development by offering the possibility that a pharmaceutical company's investment will be repaid, a powerful incentive to companies to invest millions and millions of dollars into risky research and development of these medications. Without patent protection, other manufacturers could copy new drugs immediately. Since their costs are minimal, they

can offer their versions at a reduced price, seriously hurting the ability of the company that developed the drug to recoup its costs." *www.america.gov/st/business-english/2008/April/20080429230451myle en0.4181027.html*

Closure

Questions emerging from the discussion are recorded on a class "Brainstorm" sheet to be answered as the unit proceeds.

Assessment

"We Are Family" sheets are assessed for accuracy and discussions are monitored for participation.

Name: ______________________________ Date: ____________

We Are Family

All families come in different sizes and have different relationships.

Please think of your immediate and extended family and fill out Chart 1 of this form to the best of your knowledge.

Immediate family (parents, siblings, grandparents)

Extended family (aunts, uncles, cousins, nieces, nephews)

Chart 1. Personal Data

Age group	Immediate family	Extended family	Total Family Members
0–11 years			
12–19 years			
20–59 years			
60 years or older			

We will combine our personal data into class data below in Chart 2 to determine an estimate of the approximate pharmaceutical usage for our class "family." The "Pharma Factor" is derived by adding up the percentages of people who take any number of prescription drugs in their age group. The factor will represent the percentage of people who take "1 or more" prescription drugs (the sum of the results for 1, 2, 3–4, and 5 or more).

Chart 2. Class Data

Age group	Total Family Members for Class	Pharma Factor	Total Prescription Drugs
0–11 years			
12–19 years			
20–59 years			
60 years or older			

Total pharmaceuticals for our class "family" ____________

What's in the Bottle?

Research on Pharmaceutical Composition

To the Teacher

During this lesson, students will examine empty bottles/containers of pharmaceutical compounds and drugs. They will conduct internet research on commonly prescribed and over-the-counter medications.

Objective

Students will recognize the relationship between chemical formulas and everyday medicines.

Time Needed

One 50-minute class period

Materials

- Empty containers (bottles or boxes) of pharmaceuticals with labels (can be obtained from a pharmacy or created by the teacher with information about a common medication)
- Drug Profile Sheet
- Computers with internet access

Procedure

1. Divide class into groups of three to four students. Give each group two to three empty medicine containers for various common medications and interesting drugs.

2. Instruct students to complete a Drug Profile Sheet for each container that their group has been given. They should also use the classroom or computer lab computers to find more information about the drugs. A combined drug profile library should be made by compiling all of the information from each group.

Closure

Students share interesting facts they have learned about the various medications. What are some commonalities?

Assessment

Student Drug Profile Sheets are assessed for accuracy and completeness.

Name: ______________________________ Date: ____________

Drug Profile Sheet

Please fill out one drug profile sheet per container for your group. Take as much information from the packaging as possible and then use the internet to fill in more information about the drug. Use the back of this sheet if necessary.

Drug name ______________________________

Active ingredient or chemical name ______________________________

Other components ______________________________

____

Serving size (mass? volume?) ______________________________

Indications (what is it used for?) ______________________________

Side effects ______________________________

Warnings ______________________________

Manufacturer ______________________________

These questions can be answered from sources on the internet.

Chemical formula ______________________________

Chemical structure (draw in box):

What is its history? When was it discovered?

When was it approved? Is this the only drug of its kind?

Other interesting features?

____

____

____

____

____

____

Lesson 3

How Do Drugs Work?

Molecular Models and Drug-Target Interactions

To the Teacher

In this lesson, students will learn about the basic structure of some common drugs and compounds. They will see examples of how the drugs interact with their target as well. The content focus will be on intermolecular forces, the active site and molecular structure.

Objectives

Students will be able to identify the molecular structure of some common drugs and articulate their action at target sites.

Time Needed

One 50-minute class period

Materials

- Molecular Model Kits (note that gumdrops, clay, or mini-marshmallows and toothpicks can be used)
- Computers with internet access
- "How Do Drugs Work?" handout

Procedure

1. Challenge students to create models of some common drug compounds using molecular model kits (or alternative materials). Student teams can be assigned randomly to different compounds so that there is a variety of

structures to display once completed. Some suggested compounds to use for structures include:

- acetylsalicylic acid,
- acetaminophen,
- phenobarbital, and
- desipramine.

2. Students can use the "Pub Chem" (*http://pubchem.ncbi.nlm.nih.gov*), which includes detailed information about dozens of common drugs, include chemical structures, UV spectra, apparatus, and parameters for chromatographic separation, physical properties, and activity type.
3. Have students display their models along with the name and chemical formula for other students to view.
4. Once the models are completed, show molecular modeling of drug/target interactions of several drugs. RasMol can be used or pictures and videos already available online can be shown. To download RasMol: *www.umass.edu/microbio/rasmol*
5. Now that students have seen some of the structures of the drugs and how they interact with the active site, it is important to spend some time clarifying the chemistry content that has been introduced. Introduce intermolecular forces and explain how these interactions are important at the active site:
   - Hydrogen bonding
   - Dipole-dipole
   - Dispersion forces (London forces)
   - Electrostatic forces
6. Initiate a discussion: "You have already learned about chemical bonds between atoms and now you have seen that there are also forces between molecules—intermolecular forces. Do you think that chemical bonds or intermolecular forces are used at the active site? Why?

Closure

Most of the drugs that we have looked at so far are ingested. Think about the drug going from being ingested to being used by different parts of the body. What does it need to successfully arrive at its target? What would we see in terms of solubility? Ability to withstand low pH? Do you think that all drugs are ingested in an active form or do they change once absorbed?

Assessment

Students complete "How Do Drugs Work?" sheet and are assessed on completeness and accuracy.

Name: ______________________________ Date: ____________

How Do Drugs Work?

1. **Describe in a full paragraph how a drug interacts with its target. Take information from our discussion in class, your textbook and internet sources. Make sure that all of the paragraph is in your own words and include citations if you used sources other than your textbook.**

2. **Explain why you agree or disagree with the following statement:**

   "Covalent bonds are not formed between a drug and its target at the active site. Only intermolecular forces hold the drug to the target."

Lesson 4

How Are Pharmaceuticals Made?

Creating a Synthetic Drug

To the Teacher

During this lesson, students will complete a lab activity where they will synthesize aspirin. They will then read about the development and regulation of pharmaceuticals in the United States.

Objectives

Students will be able to identify the key components of aspirin, describe its preparation, and articulate key steps in the regulatory processes surrounding pharmaceutical development and approval.

Time Needed

One 50-minute class period for lab; reading assignment for homework

Materials

- salicylic acid
- acetic acid
- concentrated H_2SO_4
- ice water
- 100 ml Erlenmeyer flask
- stirring rod
- Buchner funnel
- filter paper
- water-bath

- "Making Medicine" sheet
- "How Drugs Are Developed and Approved" reading

Procedure

1. Review chemical reactions including synthesis reactions, single displacement, double displacement, and decomposition reactions on the whiteboard or overhead.
2. Inform students that they are going to be synthesizing a medicine today, but do not tell them which one (aspirin, or acetylsalicylic acid).
3. Review the lab procedure, which is on the "Making Medicine" sheet, with the students.
4. Once students have completed the lab, conduct a class discussion about the formula and structure of the drug that they just synthesized. The students should then use their drug profile sheets to determine which substance they made.

Aspirin
Acetylsalicylic Acid
$C_9H_8O_4$

Closure

Review the steps involved in making a chemical compound. Concepts of purity and percent yield should be discussed. How do we ensure that the pharmaceuticals on the market meet safety standards? Assign homework reading on "How Drugs Are Developed and Approved" and ask students to answer questions.

Assessment

Students are assessed based on successful implementation of laboratory procedures and on completeness and accuracy of homework responses.

Name: ______________________________ Date: ____________

Making Medicine

Purpose

To synthesize a pharmaceutical from salicylic acid and acetic acid.

Equipment

100ml Erlenmeyer flask, stirring rod, Buchner funnel, filter paper, water-bath

Materials

salicylic acid, acetic acid, concentrated H_2SO_4, ice water

Procedure

1. Put 3 g of salicylic acid into a 100 ml Erlenmeyer flask.
2. Carefully add, stirring constantly, 6 ml of acetic acid, followed by 10 drops of concentrated sulfuric acid.
3. Swirl the contents of the flask.
4. Heat the flask in a boiling-water bath for 15 minutes.
5. Remove the flask from the bath.
6. While the contents are still hot, cautiously add 5 ml of ice water at once.
7. After the reaction subsides, add 35 ml of water and chill the contents of the flask in an ice bath.
8. Use a stirring rod to break up any lumps of solid that may form.
9. Collect the product by vacuum filtration using a small Buchner funnel.
10. In the funnel, rinse the product with 15 ml of ice water.
11. Transfer the solid to a sheet of dry filter paper and allow it to dry thoroughly.

Questions

1. What medicine have you made? Consult your drug profile sheets to find out.
2. Write the reaction here:

***Safety Note:* DO NOT ingest the medicine. The medicine you have made is not pure enough to be consumed.**

Reading: How Drugs Are Developed and Approved

(This article is reprinted from *www.fda.gov/Drugs/DevelopmentApprovalProcess/HowDrugsareDevelopedandApproved/default.htm*)

The mission of FDA's Center for Drug Evaluation and Research (CDER) is to ensure that drugs marketed in this country are safe and effective. CDER does not test drugs, although the Center's Office of Testing and Research does conduct limited research in the areas of drug quality, safety, and effectiveness.

CDER is the largest of FDA's five centers. It has responsibility for both prescription and nonprescription or over-the-counter (OTC) drugs. The other four FDA centers have responsibility for medical and radiological devices, food, biologics, and veterinary drugs.

Some companies submit a new drug application (NDA) to introduce a new drug product into the U.S. Market. It is the responsibility of the company seeking to market a drug to test it and submit evidence that it is safe and effective. A team of CDER physicians, statisticians, chemists, pharmacologists, and other scientists reviews the sponsor's NDA containing the data and proposed labeling.

The section below entitled *From Fish to Pharmacies: The Story of a Drug's Development,* illustrates how a drug sponsor can work with FDA's regulations and guidance information to bring a new drug to market under the NDA process.

From Fish to Pharmacies: A Story of Drug Development

Osteoporosis, a crippling disease marked by a wasting away of bone mass, affects as many as 2 million Americans, 80 percent of them women, at an expense of $13.8 billion a year, according to the National Osteoporosis Foundation. The disease may be responsible for 5 million fractures of the hip, wrist and spine in people over 50, the foundation says, and may cause 50,000 deaths. Given the pervasiveness of osteoporosis and its cost to society, experts say it is crucial to have therapy alternatives if, for example, a patient can't tolerate estrogen, the first-line treatment.

Enter the salmon, which, like humans, produces a hormone called calcitonin that helps regulate calcium and decreases bone loss. For osteoporosis patients, taking salmon calcitonin, which is 30 times more potent than that secreted by the human thyroid gland, inhibits the activity of specialized bone cells called osteoclasts that absorb bone tissue. This enables bone to retain more bone mass.

Though the calcitonin in drugs is based chemically on salmon calcitonin, it is now made synthetically in the lab in a form that copies the molecular structure of the fish gland extract. Synthetic calcitonin offers a simpler, more economical way to create large quantities of the product.

FDA approved the first drug based on salmon calcitonin in an injectable. Since then, two more drugs, one injectable and one administered through a nasal spray were approved. An oral version of salmon calcitonin is in clinical trials now. Salmon calcitonin is approved only for postmenopausal women who cannot tolerate estrogen, or for whom estrogen is not an option.

How did the developers of injectable salmon calcitonin journey "from fish to pharmacies?"

After obtaining promising data from laboratory studies, the salmon calcitonin drug developers took the next step and submitted an Investigational New Drug (IND) application to CDER.

Once the IND application is in effect, the drug sponsor of salmon calcitonin could begin their clinical trials. After a sponsor submits an IND application, it must wait 30 days before starting a clinical trial to allow FDA time to review the prospective study. If FDA finds a problem, it can order a "clinical hold" to delay an investigation, or interrupt a clinical trial if problems occur during the study.

Clinical trials are experiments that use human subjects to see whether a drug is effective, and what side effects it may cause.

The salmon calcitonin drug sponsor analyzed the clinical trials data and concluded that enough evidence existed on the drug's safety and effectiveness to meet FDA's requirements for marketing approval. The sponsor submitted a New Drug Application (NDA) with full information on manufacturing specifications, stability and bioavailablility data, method of analysis of each of the dosage forms the sponsor intends to market, packaging and labeling for both physician and consumer, and the results of any additional toxicological studies not already submitted in the Investigational New Drug application.

New drugs, like other new products, are frequently under patent protection during development. The patent protects the salmon calcitonin sponsor's investment in the drug's development by giving them the sole right to sell the drug while the patent is in effect. When the patents or other periods of exclusivity on brand-name drugs expire, manufacturers can apply to the FDA to sell generic versions.

Questions

After reading the article, please write a few paragraphs about your reaction to the article and these questions:

1. Do you think there is an appropriate amount of oversight? Why or why not?

2. What would you still like to know about the process of drug approval?

3. Do you feel that the medications that you and your family take are safe? Why or why not?

7

Lesson 5

Vioxx

A Case Study

To the Teacher

In this lesson, students will learn about the process of removing a pharmaceutical from the market by engaging in a case study of Vioxx. They will proceed through an online activity called, "Amanda's Absence: Should Vioxx Be Kept Off of the Market?" in order to grapple with the scientific, ethical, and economic considerations involved in pharmaceutical regulation.

Objectives

Students will be able to articulate the purpose and processes involved in clinical trials and will appreciate the complex scientific, ethical, and economic consequences of removing a pharmaceutical from the market.

Time Needed

Two class periods

Materials

- Computers with internet access
- "Amanda's Absence" sheet

Teacher Preparation

Review the timeline of the Vioxx case at *www.npr.org/templates/story/story.php?storyId=5470430* to familiarize yourself with the history of the case.

Procedure

1. Show students the following television commercial for Vioxx from 2004, the year it was removed from the market: *www.youtube.com/watch?v=Lel03EKzqsg*
2. Inform the students that Vioxx was a drug that was introduced to the market in 1999 and subsequently removed. Ask, "Why might a drug be removed from the market?"
3. Assign students to groups of two to three. Explain that they are going to be following an online case study about Vioxx. They should be instructed to answer all of the questions and then develop a final written recommendation.

Note: The case study was written by Dan Johnson, Department of Biology, Wake Forest University. The links are:

- "Amanda's Absence: Should Vioxx Be Kept Off the Market?" *http://sciencecases.lib.buffalo.edu/cs/collection/detail.asp?case_id=180&id=180*
- "Amanda's Absence" case study teacher link: *http://sciencecases.lib.buffalo.edu/cs/files/vioxx_notes.pdf*

Closure

Student groups share their findings and discuss their recommendations for Vioxx.

Assessment

Students are assessed on question responses and final written recommendations.

Name: ______________________________ Date: ____________

Amanda's Absence Case Study Questions[1]

Part 1

1. Some prescription drugs may remain on the market 20 years or more. Others are removed shortly after being introduced. Give three to four reasons that a drug might be removed from the market, either by the FDA or its maker. Is every reason true for every drug?

2. How might the manufacturers have determined that Vioxx increases the risk of heart attack? Based on your answers, when would they have learned this information?

3. What other facts or information might Dr. Sharpe (or you) want to know about Vioxx? About Amanda's condition?

Part 2

1. What are the advantages and disadvantages of Vioxx versus other pain-relieving medicines? Why are these important?

2. According to the data provided to FDA, are all patients taking Vioxx at greater risk of a heart attack or stroke? Why or why not?

3. What are two *other* questions that you have about the Vioxx withdrawal that were not addressed by the press release?

---

1. Questions by Dan Johnson, Department of Biology, Wake Forest University *http://sciencecases.lib.buffalo.edu/cs/collection/detail.asp?case_id=180&id=180*

Part 3

1. What are two strengths and two weaknesses in the current system of drug approval? Why did you choose these particular strengths and weaknesses?

2. Based on your answers to Question 1 above, did the FDA approve Vioxx too soon? Why or why not? Could the approval system be changed in a way that prevented the heart attack deaths attributed to Vioxx?

3. Should Merck & Co., be punished for putting an unsafe drug on the market? Why or why not?

Part 4

Please make sure to put your written recommendation on another sheet of paper and turn in with your case study questions.

Congressional Subcommittee Hearing

Should Prescription Drugs Be Banned From TV?

To the Teacher

During this three-period activity, students will research and debate whether "direct-to-consumer" (DTC) pharmaceutical ads should be banned from television. Students will be assigned to specific roles and will contribute to a report developed by a fictional congressional subcommittee to develop a bill on the issue.

Objectives

Students will be able to identify and appreciate the viewpoints of multiple stakeholders in a controversial socioscientific issue and use evidence to advance claims and support decisions.

Time Needed

Three class periods (one period for research and preparation and two periods for the hearing)

Materials

Computers with internet connection

Procedure

Day 1

1. Present students with the following scenario and assign roles:

   Rising prescription drug costs and the deluge of direct-to-consumer (DTC) ads on TV have led some to question whether pharmaceutical companies should advertise on TV. Several recent reports have suggested that pharmaceutical companies spend many times more dollars on promoting their products than producing them, and that those costs are passed on to consumers. Some have also suggested that the ads can be misleading to consumers. Advocates of DTC ads cite the First Amendment freedom of speech, as well as a business' right to make a profit, in support of the ads. They also suggest that the ads are educational.

   This controversy has come to a head and the U. S. Congress has requested a subcommittee hearing on the issue of whether Congress should develop a bill to ban DTC ads on TV. Your group has been invited to make a presentation to a congressional subcommittee explaining your position on this question. Use your materials and additional research to develop your arguments as well as a position paper on the issue. After the arguments are made for and against the bill, the subcommittee members will vote on whether to adopt each proposed change, and Congress (the class) will vote whether it will pass or not.

   You have one day to prepare with your group and two days for presentations and arguments. Each group will be expected to keep a record of the proceedings and be prepared to provide their research sources and support the reliability of their information.

   The Roles:

   - **Pharmaceutical Company Representatives** (against ban): research arguments that support freedom of speech, costs of research, importance of consumer education, responsibilities to investors, and so on.
   - **Consumers of Pharmaceutical Drugs** (may be for or against ban): research arguments about consumer choice, rights in determining one's own health care, misleading ads, drug lawsuits, and so on.
   - **American College of Physicians** (for ban): research arguments that these advertisements are misleading, costly, and negatively impact doctor-patient relationship.

- **FDA's Center for Drug Evaluation and Research (CDER)** (neutral): investigate research on the accuracy of the ads, their impact on consumers, and the ability of the FDA to adequately regulate the ads.
- **Senate Subcommittee Members:** conduct background research on both the pro-ban and anti-ban arguments so you can ask questions of the other groups at the hearing.

2. Students can familiarize themselves with the function of congressional committees and subcommittees at *http://kids.clerk.house.gov/middle-school/lesson.html?intID=34.*
3. Provide students with the remainder of the period to research their arguments and develop their opening statements.

Day 2

Provide students with the following outline of the process to be followed in the hearing:

Subcommittee Hearing on a Proposed Ban of DTC Advertising on TV
(5 minutes per presentation/questioning/rebuttal):

- Students will break into groups and discuss final preparations.
- Senate Committee Chairperson makes opening remarks
- Pharmaceutical Company Representatives present
- Committee Questioning
- Consumers of Pharmaceuticals present
- Committee Questioning
- American College of Physicians presents
- Committee Questioning
- FDA's CDER presents
- Committee Questioning
- Teacher will provide a recap and remind students that each group will have time for a rebuttal statement the next day, and can continue their research for homework.

Day 3

- Students will break into groups and discuss their rebuttal statement
- Committee Chairman will deliver a recap of the day before
- Pharmaceutical Company Rebuttal
- Consumers Rebuttal
- American College of Physicians Rebuttal
- FDA's CDER Rebuttal
- Committee Questioning
- The committee will work on their final recommendation as the other groups work on their final written statements.

Sample Resources

(can be shared with students, but given more as examples of the types of things that they will find)

- Congressional Budget Office. Potential Effects of a Ban on Direct-to-Consumer Advertising of New Prescription Drugs (*www.cbo.gov/publication/42186*)
- FDA. The Impact of DTC Advertising (*www.fda.gov/Drugs/ResourcesForYou/Consumers/ucm143562.htm*)
- FDA. How Drugs Are Developed and Approved. (*www.fda.gov/Drugs/DevelopmentApprovalProcess/HowDrugsareDevelopedandApproved/default.htm*)
- Drug Discovery and Development: Investment of Resources (*www.innovation.org/drug_discovery/objects/pdf/RD_Brochure.pdf*)
- CDER: The FDA's Consumer Watchdog for Safe and Effective Drugs (*www.fda.gov/Drugs/ResourcesForYou/Consumers/ucm143462.htm*)
- Centers for Disease Control. Prevalence of Pharmaceutical Use in United States (*www.cdc.gov/nchs/data/databriefs/db42.htm*)
- Public Citizen's Comments to the FDA (*www.citizen.org/hrg2002*)

- American Advertising Federation's Support of DTC Ads (*www.aaf.org/default.asp?id=248*)
- American College of Physicians Statement Against DTC Ads (*www.acponline.org/advocacy/current_policy_papers/direct_prescript.pdf*)
- Editorial: Take Ads Off TV. (*http://usatoday30.usatoday.com/news/opinion/editorials/2005-06-12-oppose_x.htm*)

Closure

Allow the congressional subcommittee to read their decision. Begin a class discussion to debrief the subcommittee hearing experience. What were the strongest arguments? What compromises might have to be made to be fair to particular groups? Where did you notice conflicting evidence on the same claim? What have we learned over the course of this unit?

Assessment

Students should write a reflection paper about the presentations and debate as well as the activities from the last two weeks. They should be given copies of the questions from the first day's discussion along with a copy of the "class brainstorm." Reflections are assessed based on level of detail, demonstration of understanding of key issues, and appreciation of the complexity of the controversy.

Reflection Scoring Rubric

1 pt.	2 pts.	3 pts.	4 pts.	5 pts.
Reflection fails to convey information on knowledge gained or reflect on student's understanding of the complexity of the controversy, or remaining questions.	Reflection includes some insights into student's understanding of the complexity of the controversy but fails to include any information on knowledge gained or questions remaining.	Reflection conveys information on knowledge gained, but fails to reflect on student's understanding of the complexity of the controversy.	Reflection conveys information on knowledge gained, insights into the complexity of the controversy and any remaining questions, but does so in an unclear or confused manner.	Reflection conveys information on knowledge gained, insights into the complexity of the controversy and any remaining questions in a clear and thoughtful manner.

Name: ______________________________ Date: ______________

Notes From Congressional Subcommittee Hearing

	Opening Statement Key Points	Rebuttal Key Points	Subcommittee Questions	My Questions/Notes
Pharmaceutical Company				
Consumers				
American College of Physicians				
FDA				

A Final Word (for Now)

We have presented many national and international talks and workshops on the socioscientific issues framework and the important role it plays in science classrooms. We have found many forward-thinking teachers who are excited to have found a strategy that can engage and positively impact their students' educative experiences. And we have also heard many teachers enthusiastically ask "Where can I get my hands on some of these materials?"

This book is aimed at fulfilling that reasonable request. But we do so with a cautious sense of optimism because it is apt to be "too helpful" and possibly misleading to some individuals. We have stressed in the book that there is no one template to follow for incorporating SSI in your classroom. If one methodically attempts to follow a rigid recipe, the results will likely fall short of their mark. The best educational materials can be made infertile by infertile minds. However, your own ingenuity, creativity, knowledge of subject matter, and developmental understanding of your students will become invaluable intellectual resources to you as you use the SSI framework and the examples within this book as a guide to organize your teaching strategies and curriculum.

If you have made it to this point in the book, you likely exhibit the virtue of perseverance—a virtue necessary for all others that may follow! Moral excellence depends on doing things repeatedly well. We are confident that your sense of professionalism, your commitment to engaging children in the activity of science, and your concern for developing functional scientific literacy that will add to a public understanding of science will enable you to seek novel ways to use the ideas and resources offered by this book to accomplish these critical goals of science education.

—Dana Zeidler and Sami Kahn, 2014

Index

*Page numbers printed in **boldface** type refer to tables or figures.*

N

O

P